Responsible Corporate Strategy in Construction and Engineering

The role that communities have to play in the evolution and implementation of an effective strategy is often overlooked, despite the fact that it is becoming increasingly important and complex. This book addresses how consulting and contracting firms in the construction and engineering industries integrate corporate social responsibility (CSR) into business strategy and how this translates into better business performance. In the context of the current global drive towards sustainability, it seeks to untangle the rhetoric and reality of CSR, providing a guide to effective and meaningful engagement with the community in the boardroom.

Ideas, concepts, theories and debates in the previously separate areas of corporate performance, corporate social responsibility, corporate strategy and corporate governance are integrated for the first time in this book, promoting a more liberal and wider debate about CSR. The result is an important and timely examination of a new challenge which faces every firm in the engineering and construction industries, from the very largest transnational corporations to consultants, and to the many thousands of small and medium-sized enterprises that employ the vast majority of people in the industry.

Contemporary research is integrated with practice throughout this book in a clear and easy-to-follow style. The extensive use of real-life examples ensures that the content is of value to managers who have to deal with the reality of the industry on a day-to-day basis. The case studies of some of the world's leading firms provide rich qualitative data to support the arguments and are an excellent source of learning and teaching material to the subject of CSR for students.

Martin Loosemore is Professor of Construction Management in the Faculty of the Built Environment at the University of New South Wales, Australia. He has published many articles and books on risk management, project management and human resource management.

Florence Phua is a Reader in the School of Construction Management and Engineering at the University of Reading, UK. She specialises in international construction management and publishes widely in the area.

Responsible Corporate Strategy in Construction and Engineering

'Doing the right thing?'

Martin Loosemore and Florence Phua

Spon Press
an imprint of Taylor & Francis
LONDON AND NEW YORK

First published 2011
by Spon Press
2 Park Square, Milton Park, Abingdon, Oxon OX14 4RN

Simultaneously published in the USA and Canada by Spon Press
270 Madison Avenue, New York, NY 10016, USA

Spon Press is an imprint of the Taylor & Francis Group, an informa business

Typeset in Sabon by
GreenGate Publishing Services, Tonbridge, Kent
Printed and bound in Great Britain by
CPI Antony Rowe, Chippenham, Wiltshire

British Library Cataloguing in Publication Data
A catalogue record for this book is available from the British Library

Library of Congress Cataloging in Publication Data
Loosemore, Martin, 1962–
Responsible corporate strategy in construction and engineering: "doing the right thing?" / Martin Loosemore and Florence Phua.
p. cm.
Includes bibliographical references.
1. Construction industry—Management. 2. Building—Moral and ethical aspects. 3. Social responsibility of business. 4. Customer relations. 5. Consumer satisfaction. 6. Strategic planning. I. Phua, Florence, 1974– II. Title.
HD9715.A2L657 2010
624.068'4—dc22
2009051835

ISBN: 978-0-415-45909-9 (hbk)
ISBN: 978-0-415-45910-5 (pbk)

Contents

List of figures

List of tables

Preface

This book addresses how consulting and contracting firms in the construction and engineering industries integrate CSR into business strategy and how this could translate into better business performance. It challenges the simplistic and one-sided reporting the subject often receives and seeks to untangle the rhetoric and reality of CSR through balanced and reasoned arguments from both sides of the debate.

Responsible Corporate Strategy in Construction and Engineering is written for a new generation of progressive leaders who recognise that in the future competitive advantage will be achieved by leveraging human, intellectual and social capital as well as monetary and physical capital. It is written for those managers who seek to better align the economic imperatives of shareholders and owners with a sense of social responsibility to other key stakeholders in their business. These are new challenges which face every firm in the engineering and construction industries, from the very largest transnational corporations to the many thousands of small and medium-sized enterprises that employ the vast majority of people in the industry.

This book, for the first time, integrates ideas, concepts, theories and debates in the previously separated areas of corporate performance, corporate social responsibility, corporate strategy and corporate governance. In doing so, it offers a more in-depth and objective debate about CSR by drawing perspectives from current research in the humanities, social sciences, economics, politics and philosophy.

While this book draws upon contemporary research, it also integrates with practice in a clear and easy-to-follow style. Real-life examples are used extensively throughout the book and nine in-depth case studies of some of the world's leading firms provide rich qualitative data to support our arguments.

The key features of this book which distinguish it are:

- The integration of contemporary theory, research and practice into a coherent model of social corporate strategy and governance.
- A liberal and in-depth approach which draws from a variety of disciplines in business and management, humanities, and social and physical sciences.
- A global perspective which uses examples from companies operating across international boundaries.
- A cross-sector approach which draws examples from a variety of industries.
- An extensive use of real-life case studies and examples of principles in practice.

Acknowledgements

The authors wish to express their gratitude to the following people who kindly gave their time for interviews and case studies or valuable feedback on various drafts of this book.

Maria Atkinson, *Global Head of Sustainability*, Lend Lease Corporation

Cathy Crawley, *Principal*, Arup Sustainability, Arup

David Hughes, *Director*, Crown Project Services

James Kell, *Chairman and Chief Executive Officer*, Kell and Rigby Ltd

Ken Maher, *Chairman*, HASSELL

Tony McCormick, *Principal*, HASSELL

Michael Myers, *Managing Director*, CONCENTRIC Asia Pacific Pty Ltd

Dr Caroline Noller, *Head of Corporate Responsibility*, The GPT Group

Liz Potter, GM Global Sustainability Group, Lend Lease Corporation

Dr Tony Stapledon, *Group Sustainability Manager*, Leighton Contractors Pty Ltd

Siobhan Toohill, *General Manager*, Corporate Responsibility & Sustainability, Stockland

1 Corporate social responsibility

In this chapter we discuss the role of firms as agents for social good. We explore contemporary debates about corporate social responsibility (CSR) and provide a framework within which judgements can be made about appropriate policies and practices. Our discussion is placed in the context of the construction and engineering industries at a domestic and global level and our aim is to provide a balanced perspective which addresses both the potential benefits and significant rhetoric surrounding the CSR debate.

Introduction

In very simple terms, the CSR movement has arisen out of six fundamental questions:

1. What is the relationship between organisations and the communities in which they operate?
2. What responsibilities do businesses have to society?
3. What role should and can firms play in being agents for positive social change?
4. Do firms have responsibilities which go beyond the benefits their economic success already brings to society?
5. Do firms have greater rights if they accept greater social responsibilities?
6. Is there a conflict between a firm's relationship with its shareholders and with its stakeholders?

The purpose of this chapter is to attempt to answer these questions, presenting both sides of the argument so that readers can make up their own minds. In doing so, we are cognisant that different cultures, nations and societies define the relationship between business and society differently. For example, people in wealthy societies generally expect higher wages and better working conditions than people in poorer societies, where it is more likely that the basic necessities of life will be the primary concern. Similarly, attitudes towards gender and equality in the workforce can vary considerably between different societies. This issue of 'cultural relativity' represents one of many emotive dilemmas in the CSR debate which we discuss later in this book. The point we make here is that judgements about right and wrong and, to a greater extent the role and primary functions of a firm, are inevitably judged from a cultural and societal context.

Defining CSR

The somewhat confusing nomenclature which surrounds CSR is a challenge for anyone wishing to come to grips with the subject. To avoid any confusion, we define the concept of CSR from philosophical, conceptual and practical perspectives.

Philosophical definitions of CSR

The most fundamental issue from which all debate on CSR arises is whether or not corporations, on the basis of economic grounds, should be socially responsible for the betterment of societies and if so, why, and how? This debate was polarised when Nobel Prize winning economist Milton Friedman (1962) argued that the only social responsibility of firms is to grow profits and maximise the wealth of shareholders. By doing this, firms are able to maximise social and economic advantages through the provision of employment, security and wages etc. According to Friedman (p.122), the main goal of a firm is to serve the interests of its owners by making 'as much money as possible while conforming to the basic rules of the society, both those embodied in law and those embodied in ethical custom'. Friedman chose to interpret social issues to mean non-business issues, enabling him to dismiss the idea of CSR and claim that the 'business of business is business'. Indeed, he argued that promoting CSR would positively damage business interests and even foster the advancement of socialism which was the great fear of the time in the USA. While rhetorical in parts, Friedman's formative article raised a whole series of important questions at the time. Nevertheless, today the assumptions of neoclassical economics which underpinned his work are increasingly being questioned, particularly the assumed positive relationship between profit and social outcomes and the role of markets in automatically generating social well-being (Pearce 2006; Macfarlane 2008). The recent global financial crisis has exposed the fragility of many assumptions which underlie rational market theory and it is increasingly accepted that effective economic systems and free markets depend on socially responsible firms engaging in transparent, honest and fair transactions (Vranceanu 2007; Washington 2009). As we tackle contemporary business challenges like globalisation and climate change, it is becoming apparent that the central neoclassical economic concept of 'utility' (the ability of goods or services to satisfy a consumer's needs) requires broader definition than simply in terms of economic capital. Other forms of capital which have become more relevant in recent times include: manmade capital (roads, health services, ports and airports, etc.); natural capital (clean water, managed forests, unpolluted oceans, healthy ecosystems and atmosphere, etc.); social capital (human relationships, vibrant and close knit communities, etc.); cultural capital (traditions, value and belief systems); and human capital (access to knowledge and skills, etc.). This raises two fundamental questions which are central to the CSR debate for all businesses including those in construction and engineering: how do you measure and compensate society for the full impact of business activities?; and what does the term 'social' actually mean for business from a strategic viewpoint?

In addressing the first question of how to measure and compensate society for the full impact of firms' activities, market solutions to the most pressing depreciations (like climate) are starting to emerge. These solutions, which essentially commoditise industrial externalities (for example, carbon) include initiatives such as the EU Emissions Trading Scheme and Energy Performance of Buildings Directive and Australia's proposed Carbon Pollution Reduction Scheme and the National Greenhouse and Energy Reporting Act 2007. Yet the risks and opportunities associated with these new initiatives are not well understood in the construction and engineering industries, particularly by the many small to medium-sized enterprises (SMEs) that dominate it. Despite impressive reports of improved environmental performance on selected construction and engineering projects, we show in this book that many construction and engineering firms, building owners and clients remain either unprepared for these new challenges or do not see them as a business imperative. This means that claims of apparent advances in CSR within the industry can be misleading at best and at worst, obscure underlying trends which may be going in the opposite direction (RICS 2008). Nevertheless, recent research does indicate that CSR is increasingly becoming a strategic focus for CEOs, with evidence suggesting that market leaders in areas such as climate change can generate up to 80 per cent extra value for their business, while market followers can suffer up to 65 per cent value-at-risk (Beatty 2009). Although issues such as climate change and globalisation may seem irrelevant to the many SMEs that dominate the construction and engineering industries, their implications will sooner than later filter down supply chains to affect their business, forcing changed practices throughout the whole industry.

In addressing the second question about the meaning of 'social' from a strategic business perspective, it is increasingly accepted that it is necessary to distinguish between stakeholder issues and wider social issues. As we will see in this book, a bewildering array of causes have been labelled as 'social' in the CSR literature making it almost impossible to define what constitutes a 'social' issue from a business perspective. The boundaries of CSR remain fuzzy for many firms in the construction and engineering industries, which in part explains why progress in promoting CSR has been slow. To be strategic, managers and corporations first need to define what constitutes a social issue for the particular organisation which at a deeper level also requires them to address the fundamental question of what a firm's responsibilities to society are. Indeed, Mintzberg (1994) even questioned the appropriateness of the term 'responsibility' when talking about business, arguing that it reflects the traditional patronising view of business as doing good 'to' society through assuming to know what is best for it. This position is echoed by Green's (2009) analysis of CSR discourse in the UK's construction industry which highlights contradictions between widely espoused social values portrayed in recent UK construction industry reform initiatives such as 'Respect for People' and the enterprise cultures (inspired by the likes of Milton Friedman) that still dominate the industry. Green argues that consequently many CSR initiatives in the UK construction industry are primarily viewed from a profit-making/business case rather than from the perspectives of those they are intended to benefit. These types of arguments are also echoed in the fields of business and management. Here, empirical research designed to establish statistically significant relationships between a firm's CSR record and its financial performance have been widely criticised. For example, Margolis and Walsh (2003) point out that such research is based on questionable assumptions that there is a measurable instrumental relationship between CSR and efficiency, that CSR can be measured using similar indicators to those used to measure financial performance and that when CSR does not seem to 'pay' then it is non-viable from a business, strategic and governance perspective. It is argued that this type of research reinforces the view that CSR is only viable if there is a business case to be made, when in reality the benefits are situated at the level of values which produces benefits (if any) that are difficult if not impossible to measure. In summary, it is broadly accepted that in philosophical terms, CSR is about achieving a balanced and sustainable path between pure capitalism (which puts profits before people) on one end of the spectrum and pure socialism (which put people before profit) on the other (Werther and Chandler 2006). Any CSR strategy should therefore be founded on an understanding that corporations serve a broader range of human values than can be captured by a sole focus on economic values.

Conceptual definitions of CSR

In conceptual terms, CSR is best explained by referring to the hierarchy described by Carroll (1991). Illustrated in Figure 1.1, it reflects the widely accepted notion that any firm should ideally satisfy four levels of responsibility to society. This starts with a firm's economic responsibility to its shareholders to provide an acceptable rate of return for their investment and is followed by, a duty to act within the legal framework drawn up by the government and judiciary, an ethical responsibility to do no harm to its stakeholders and finally, a discretionary responsibility to go further than basic requirements of shareholders, laws and ethics. It is at the discretionary

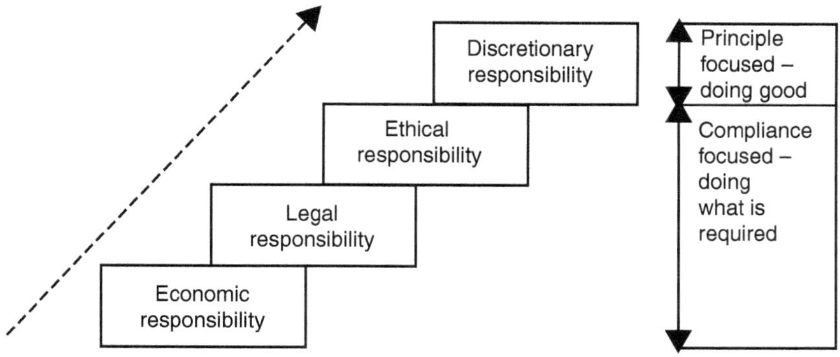

Figure 1.1 The hierarchy of CSR

Source: Adapted from Carroll 1991

level that CSR is deemed to have the most potential to add strategic value to a company, and it is where proponents would argue it has the greatest long-term positive impact. At this level, CSR clearly involves a lot more than simply adopting the latest national or international standards, although in many firms, including those in construction and engineering, this is as far as it goes (Saha and Darnton 2005; Petrovic-Lazarevic 2008; Louisot 2009). Philips (2009) argues that companies which adopt this 'box-ticking' approach to CSR have a narrow understanding of risk and are potentially vulnerable to damage to their reputation from undermined shareholder, public, employee, investor and regulator confidence.

One reason why many firms adopt a compliance-based approach to CSR is related to the need to invest in the interests of stakeholders that might not be regarded as 'primary' by all key decision-makers and uncertainty about what kinds of activities are most likely to serve their interests (Campbell 2008). To understand this further, we need to move beyond Carroll's hierarchical framework which presents a rather static view of CSR. For example, new theoretical developments founded in the natural sciences, may help to explain the potential benefits which discretionary CSR can provide to firms. Particularly innovative is Frederick's naturological model of community–business relations that treats firms as biological organisms which exist in a symbiotic relationship with a natural ecosystem of stakeholders whose survival are intimately tied to each other (Frederick 1998). In Frederick's naturological model there are two kinds of resources that ensure survival in any biological system: *income* and *capital*. Income meets an organism's immediate survival needs which can be stored (as capital) in times of abundance to see organisations through harder times. For example, in the natural world, animals living in seasonal climates stockpile food and hibernate. In the human world, people's bank savings perform a similar function and in the business world, investments do the same. Frederick argues that in biological systems an organism cannot survive if they are 'greedy' and stockpile too much at the expense of other organisms because the other organisms will rise up against the greedy to enhance their own survival. Alternatively, the greedy will perish if other like organisms die because they will become more vulnerable to predators. Thus to prevent death by predation or rebellion, it makes sense for the strong to share a portion of their income. The key point being made by Frederick is that socially responsible business behaviour is not an altruistic charitable activity but an effective and necessary strategy to ensure survival in a chaotic, competitive and ever-changing environment.

Practical definitions of CSR

The World Business Council for Sustainable Development defined CSR as a 'continuing commitment by business to contribute to economic development while improving the quality of life of the workforce and their families as well as of the community and society at large' (WBCSD 1999:3). In practical terms, this embraces a range of economic, legal, ethical and discretionary actions that enable a firm to contribute back to those economies, individuals, societies and environments in which they conduct their business activities. CSR requires that a company becomes conscious about and able to effectively manage the impact of its business decisions on people, the environment and the economy. This is a challenge which will require a response at many levels, not least in the future education of construction and engineering professionals, which some argue would benefit from a much more liberal knowledge-base than is currently the case (Beder 1998). At a firm's level, this requires commitment to social, economic and ecological objectives (the so-called triple bottom line) and an intimate understanding of the strategic relationships between them (Elkington 1999; Oury 2007). For example, from a technical and cost perspective it may be a sub-optimal solution to build a road around an environmentally sensitive area but a very good solution when one considers the lower environmental, social and reputational costs it may incur. Alternatively, from a purely financial perspective it might make no business sense to improve working conditions above construction industry norms and invest in apprenticeships, R&D, scholarships and training programmes, but good business if one considers the extra commitment, loyalty and increased retention rates from existing and new employees in times of skill shortages. The relationships between the social, economic and environmental impacts of construction and engineering activities are complex and not well understood but Figure 1.2 attempts to explore how some of these relationships may overlap.

Another practical perspective on CSR is to consider the issue of risk. As we illustrate later in this chapter, the CSR movement has, at the most basic level, largely evolved in response to the community's heightened perceptions of business as a risk to their health, safety and well-being. From this perspective, Smyth (2006) argues that a company moves through four main

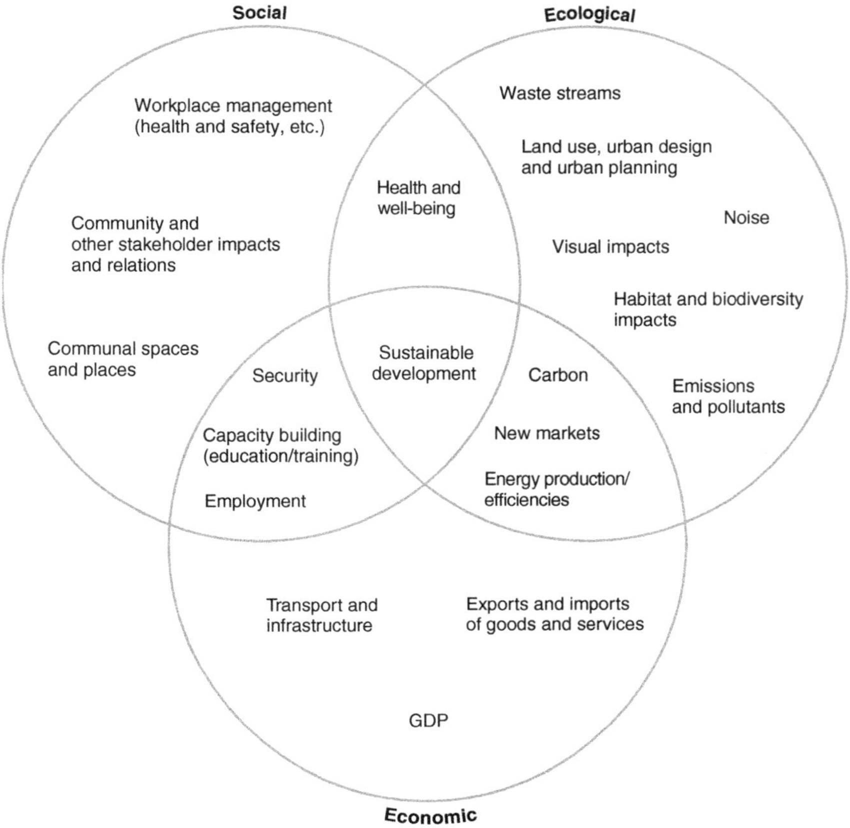

Figure 1.2 Environmental impacts of the built environment

Source: Adapted from SOTA 2009

phases of risk management maturity over time as illustrated in Figure 1.3. By combining Smyth's ideas with models of risk management maturity (such as Loosemore *et al.*'s 2005) it is possible to speculate about the typical characteristics organisations may exhibit in each phase of CSR development. For example, Smyth's ignorance phase is typically characterised by an unstructured, reactive and inward-looking approach to CSR. The focus is very much on satisfying economic responsibilities to shareholders and there is little if any consultation with outsiders. Instead outsiders are treated with suspicion and there is little if any awareness of the social, cultural, ecological and economic impacts of activities and a reactive approach to managing these risks. This level of risk management maturity is equivalent to the economic level of responsibility in Carroll's (1991) hierarchy depicted in Figure 1.1.

In Smyth's hardware phase there is some concern with these impacts and an ad hoc approach to managing these risks. There is limited experimentation with CSR initiatives on a small number of projects and by a

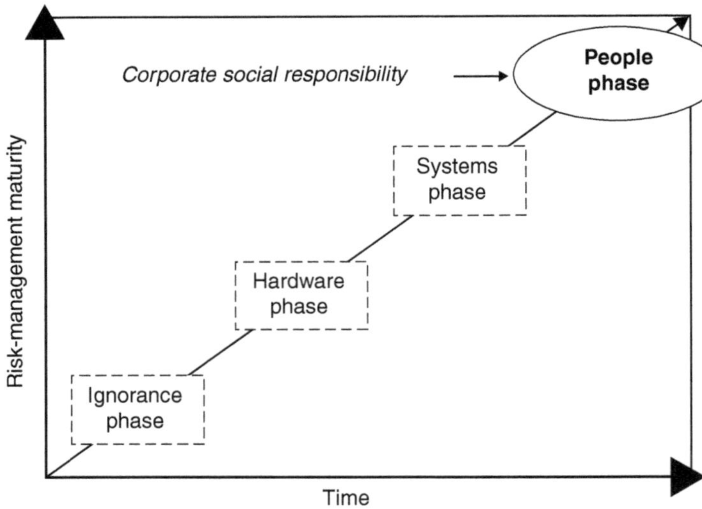

Figure 1.3 The phases of CSR maturity from a risk-management perspective

Source: Adapted from Smyth 2006

small number of people. There is also very limited consultation with the outside world and any consultation is not meaningful and is undertaken as a token gesture. CSR is generally viewed as peripheral to core business, an additional overhead which does not contribute to corporate goals, and as a general burden. CSR is not championed by senior management and there are no consistent and systematic processes in place to manage the CSR function. The focus is very much on satisfying legal and economic responsibilities and this level of risk management maturity is equivalent to the legal level of responsibility in Carroll's (1991) hierarchy depicted in Figure 1.1.

Smyth's systems phase is characterised by a compliance driven approach to CSR and by the need to satisfy economic responsibilities to shareholders, to comply with laws and regulations and with ethical codes of conduct. There are consistent processes in place to manage economic, social, cultural and ecological risks and there is systematic consultation with internal and external stakeholders. CSR initiatives are practiced on most projects and in most parts of the business and the impacts of business activities are understood. There is also a clear CSR policy with dedicated resources which is championed by senior managers and this level of risk management maturity is equivalent to the ethical level of responsibility in Carroll's (1991) hierarchy depicted in Figure 1.1.

Finally, Smyth's people phase is the challenging realm of true beyond-compliance CSR. There is collective responsibility for the full social, economic, cultural and ecological impacts of a firm's activities throughout its supply chain. For example, a construction firm would not use suppliers of natural stone if they source their raw materials from the many quarries in India and China which use child labour (Verhoef 2007). Furthermore, they are open, transparent and honest about this in their communications with business partners and outsiders. In other words, there is a CSR culture integrated into the business and its entire supply chain so that people think about the triple bottom line (economic, social and environmental) impacts of their activities instinctively every time they make a decision. CSR initiatives are also directly aligned with business strategy and the firm engages in discretionary activities which benefit society as well as business. These initiatives go well beyond the need for compliance with economic, legal and ethical responsibilities laid down by law and codes of practices. This level of risk management maturity is equivalent to the discretionary level of responsibility in Carroll's (1991) hierarchy depicted in Figure 1.1.

The CSR tipping point

The key themes that emerge from the above discussion are that CSR involves going beyond minimum requirements set down by law and having a genuine commitment through corporate policies, strategies and actions to achieve social performance as well as economic performance. It is also evident that CSR, by definition, is a discretionary activity which cannot be enforced in any court of law or regulated. Finally, CSR engagement is an evolutionary process, meaning that not all firms are ready to jump straight into a fully articulated CSR strategy. The appropriate timing will vary from firm to firm depending on the level of the maturity in its market (customer demand), its internal capabilities (employees, skills, knowledge, systems, culture, etc.) and its business environment (competitors, supply chain capability, etc.). These internal and external factors must be understood since they determine the return on investment in CSR and the point at which the firm, its supply chain and, indeed, its industry sector become receptive to CSR. This point can be referred to as the CSR 'tipping point' and is important for a number of reasons. First, the relative proximity of a firm to this point explains why some companies feel ready to engage in CSR and others do not. Second, it is likely that those firms which are most aware of their proximity to this point and take advantage of it, will benefit most because they can position themselves as markets leaders and achieve greatest competitive advantage from 'first mover' status and product differentiation. Depending on the type of products or services that firms provide, it is also probably true to say that there is a 'cut-off point' past which, a delay in implementing CSR could result in a backlash against a company from not being seen to be genuinely committed to CSR. Some industries such as the food and apparel industries have already passed the CSR tipping point, where it is unacceptable not to be seen to be investing in CSR. Whether the tipping point in the construction industry has been reached is unclear but all the discussions presented in this chapter and evidence presented in later chapters suggest that this point might be approaching faster than many firms think.

The business case for CSR

The relevance of CSR as a business strategy largely hinges on its relative costs and benefits, and ultimately its connection with performance and competitiveness. In practice, particularly, these relationships are not well understood and are often obscured by the value-laden, emotional and rhetorical nature of the CSR debate. Today, the corporate world is under increasing political pressure to be more socially responsible and the temptation is that firms in the construction and engineering industries uncritically copy and adopt CSR solutions without understanding the associated risks and opportunities to their particular business. Conversely, there is the temptation for businesses to dismiss CSR altogether or engage, as many firms have done, in superficial attempts to buy a good CSR image (Saha and Darnton 2005). Clearly, in an industry such as construction which comprises a wide variety of firms, large and small, producing many different products and services, under very different resource constraints and strategic time horizons, to assume there is one best CSR solution for all would be naïve. By extension of this argument, CSR strategy will not only vary from firm to firm but also from country to country. This view is supported by a recent survey (Table 1.1) which asked firms to rank key CSR issues in order of importance. The results show that the rankings of firms in the USA were quite different to those in Brazil or China, reinforcing the fact that a particular CSR approach might work well in one country but not in others (Economist 2008a). This is not to suggest that firms need to reinvent the CSR wheel each time they operate overseas, but rather to be aware that the business and cultural environment in which they will be working may require an adaptation to the CSR business case and its subsequent strategy.

The argued advantages of CSR

Advocates of CSR will often point to the risks of ignoring what they argue is an inevitable and irreversible trend towards greater corporate accountability and responsibility. Indeed, to the uncritical mind the evidence is quite compelling. At the time of writing this book we are experiencing a surge in shareholder and public activism as corporations collapse and victims of corporate crime are left unemployed and penniless, while executives are paid huge salary packages (Sexton 2009a). In construction, this is vividly illustrated by the ongoing scandal of James Hardie, an industrial building materials company whose scheme to avoid liabilities for asbestos compensation caused by the manufacture and use of its building products was exposed in 2004. As Sexton (2009b) points out, the name James Hardie has become a byword for corporate immorality and recent successful legal action by the Australian Securities and Investments Commission (ASIC) confirmed that the board made misleading, illegal and unethical public statements to shareholders about the firm's ability and intention to meet its compensation

Table 1.1 CSR issues in order of importance

Global rank	*Issue*	*United States*	*Britain*	*Germany*	*China*	*Brazil*
1	The environment	2	1	2	2	1
2	Safer products	5	4	6	3	2
3	Retirement benefits	4	2	1	4	7
4	Health-care benefits	1	5	8	1	8
5	Affordable products	6	3	3	5	3
6	Human-rights standards	8	8	9	9	4
7	Workplace conditions	9	10	4	7	6
8	Job losses from outsourcing	3	6	5	13	13
9	Privacy and data security	7	7	7	6	10
10	Ethically produced products	10	9	10	8	9
11	Investment in developing countries	16	11	14	12	5
12	Ethical advertising and marketing	12	12	16	11	11
13	Political influence	11	14	12	14	14
14	Executive pay	15	16	11	10	15
15	Other	13	13	15	16	12
16	Opposition to freer trade	14	15	13	15	16

Source: Going Global, *The Economist*, 17 January 2008

liabilities. Scrutiny of minutes from board meetings reveal that its relocation from Australia to the Netherlands was not only motivated by the avoidance of compensation claims but by an attempt to reduce its tax liabilities. The consequences for the ten implicated directors have been severe damage to reputation, bans imposed by the Supreme Court from future and current company directorships and heavy fines.

Since economic downturns tend to accentuate failings in corporate governance we can certainly expect increased activism in the future as shareholders and stakeholders take a greater interest in how companies are run. Few would doubt that consumer and government attitudes are changing and requiring firms to be more socially responsible. This emerging business consciousness in society means that strong stakeholder relationships have growing potential to differentiate a business through strengthened community trust in its activities, improved reputation, brand image, staff recruitment and sales of products and services (Vilanova *et al.* 2008). Certainly, the image of the construction sector could benefit from improvement. As Moodley and Preece (2009) point out, despite numerous reports which have highlighted the positive impact of the industry's activities on society and despite initiatives such as The Considerate Constructors Scheme in the UK, the public image of the construction industry is poor. Indeed, controversially, several Royal Commissions in Australia have claimed that lawless and dishonest behaviour is institutionalised within the industry (Giles 1992; Cole 2003). But the construction industry is not alone in its poor reputation, with the public's trust in all business having been eroded severely in recent times due to numerous reports of environmental pollution, white collar crime, corporate fraud and executive excesses on a massive scale. Many of the firms implicated have been leading international corporations such as BP, Enron, Arthur Anderson, One.Tel, KPMG, Morgan and Stanley, Pan Pharmaceuticals, James Hardie, Berkshire Hathaway, Lehman Brothers and AIG, to name but a few. So it is not surprising that a recent survey in Australia found that NGOs (25 per cent) have now claimed the title of most trusted institution in the public's eye, with Government slipping to 21 per cent, media at 14 per cent and business at 11 per cent (Edelman 2007b). The survey also claimed significant economic benefits from CSR practices, since 79 per cent of respondents said they would pay more for products or services from a company that was committed to doing good. This market share advantage of CSR has been proposed in many other studies such as Hinkley (2002) which found that 84 per cent of Americans said they would be willing to switch brands to a company associated with a good cause if price and quality were similar. More recently, BCA's (2007) meta-analysis of over 85,000 people in over 40 countries indicated that 90 per cent of people believe a firm should have social and environmental responsibilities beyond simply profit and that 60 per cent take this into account when choosing what brand to buy. While there is little research regarding the purchasing behaviour of construction clients in assessing tenders, it is reasonable to assume that these types of general purchasing attitudes will work through to their decisions. For example, in Sydney there is evidence that building clients are increasingly committing to improving their carbon footprints, with Green Building Council of Australia's Green Star registered projects increasing from 270 in June 2007 to 680 in June 2008. There is also evidence that commercial tenants are increasingly recognising the benefits of green buildings. Nevertheless, empirical evidence of the commercial benefits of green buildings (in terms of increased worker productivity, retention, satisfaction, etc.) is still scant as is understanding of the trade-offs different types of clients are willing and able to make against potentially increased costs.

In addition to improved image, brand and reputation, many other benefits have been linked to CSR. These include reduced operating costs, reduced insurance premiums, avoiding regulatory non-compliance, greater political clout, improved risk awareness, stronger stakeholder engagement, better community relationships and goodwill, avoidance of activism, increased workforce cohesion, attraction and retention of talented employees, strengthened partnerships with business partners, and enhanced employee well-being, health and safety and satisfaction, etc. For example, a recent study found that 70 per cent of final year undergraduate students considered a company's ethical record as crucial when choosing an employer (Oury 2007). Similarly, Kotler and Lee (2005) point to research which found that 55 per cent of MBA students would accept lower wages to work for a company that believed in something (these also tended to be the higher achieving students), that 76 per cent of these MBA students would be more likely to stay in a job that was related to a cause, and that 78 per cent are more likely to buy a product associated with a cause. Finally, there is also an apparent increased appeal to the growing number of socially responsible investors who want the longer-term stability that socially responsible investments provide (Carter and Lorsch 2004; Macfarlane 2008). As Vilanova *et al.* (2008) point out, in contrast to traditional methods for valuing companies (ratio analysis, discounted cash flow, etc.), financial analysts are increasingly using qualitative measurements to value companies considering issues such as corporate governance, core competencies, stakeholder engagement, potential for partnership and reputation, etc. Evidence to support such assertions can be found in Europe where the total volume of socially responsible investments (SRI) has increased by 36 per cent from 11.1 billion in 1999 to 15.1 billion in 2007, and in the USA where SRI account for over US$10,000 billion (over 10 per cent of all investments). Opportunities presented by evidence like this have provided significant inroads for managers wanting to build a strong business case for CSR which goes beyond short-term financial incentives. Indeed, Figure 1.4 shows that the main business benefits of CSR often cited by companies have little to do with generating higher profits. So there is still some way to go in changing perceptions about the business case for CSR.

The rhetoric of CSR

There are endless examples, such as those presented above, to support the business case for CSR. However, there is a crucial caveat attached to this attraction, which many advocates of CSR dangerously neglect: that the benefits of CSR are contingent upon a range of contextual factors that are present at the transnational (supra), country (macro), industry, organisational (meso) and even individual (micro) level. These factors dynamically, if not uniquely affect and influence how a firm operates and runs its business. This suggests that any sweeping strategic advocacy of CSR's business case is naïve, misleading, overly simplistic and, ironically, potentially damaging to its cause. CSR is certainly not risk free. For example, ACCA (2003) pointed out that there are significant potential costs associated with CSR for all stakeholders involved. These include: more meetings, briefings and policy issues for company directors; increased reporting and disclosure costs for shareholders; increased training for managers and employees; greater

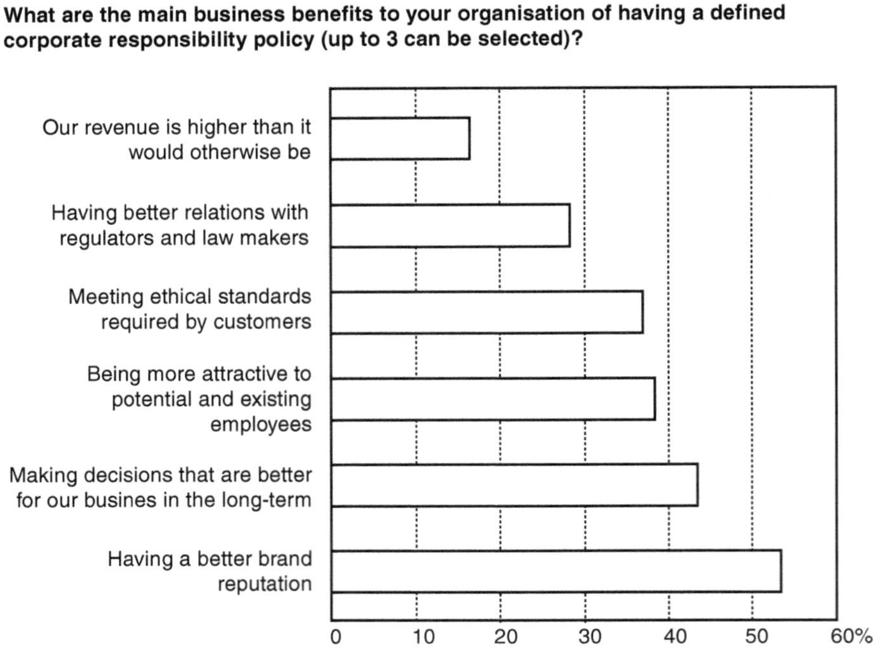

Figure 1.4 The business case for CSR

Source: A stitch in time, *The Economist*, 17 January 2008

interaction costs for communities and managers; greater short-term input costs for subcontractors; and suppliers and higher prices in the short-term for customers. Also, greater disclosure does not necessarily mean greater clarity about a firm's CSR performance. Despite decades of governments and regulators seeking to empower consumers by requiring companies to better disclose their performance, Williams (2009) argues that the resulting complexity of documentation has ironically made it easier for firms to conceal the truth about the impact of their activities.

Perhaps the biggest reservation about CSR is that the argued connection between CSR and business performance is tenuous and as yet unproven. According to Vilanova *et al.* (2008) the nature of this relationship for any firm is dependent on how the four paradoxes of CSR are managed, namely: *the strategy paradox*; *the stakeholder paradox*; *the accountability paradox;* and *the competitiveness paradox*. The strategy paradox is that the broader a firm's corporate mission and vision, the easier it is to incorporate CSR but the more difficult and impractical it is to manage and measure its impact. The stakeholder paradox is that increasing the number of stakeholders effectively reduces the capacity to control and manage the stakeholder consultation and decision-making process. The accountability paradox is that the more a company aims to be transparent and open with its stakeholders, the less able it is to communicate a coherent and central message about its vision. Finally, the competitiveness paradox is that embracing CSR can reduce certain competitive advantages, although it could strengthen other competitive factors. So, in other words, adopting CSR has the potential to create internal conflicts around new goals, values and processes which may conflict with deeply ingrained historical values and established goals and practices.

Although there is no empirical evidence relating to the construction and engineering industries, the evidence between CSR and positive performance in other industries is highly variable with some studies showing a positive relationship, some showing a negative relationship, some showing U-shaped relationships and others showing no relationships at all (Margolis and Walsh 2003; Werther and Chandler 2006; Lopez *et al.* 2007; Barnett 2008). Margolis and Walsh (2003) argue that the main problem with research that has tried to identify a relationship between CSR and business performance is that it is based on the questionable assumption that there is a *measurable* relationship to be found. They argue that research based on such an assumption forces researchers to rely on empirical data, where in reality the problem is best explored at the levels of values which are difficult to measure. Another layer of complexity is the unknown direction of any cause-and-effect relationship that is found to exist. That is to say: can CSR bring about profit to companies?; or, can companies which are profitable consequently afford to invest more in CSR and see its successful implementation?

Heugens and Dentchev (2007) also caution about the emotion-driven tendency to adopt CSR because it seems to be the right thing to do. They argue that if not managed strategically, CSR can encourage non-productive spending that encourages free-rider behaviour which can stretch organisational stakeholders beyond a manageable set and create difficulties in measuring organisational success. Ironically, it can also result in damage to reputation whereby firms lose legitimacy by associating themselves publically with too many causes which they then fail to satisfy. For example, Saha and Darnton's (2005) analysis of the sustainability practices of a sample of local and multinational companies that portray strong CSR credentials, revealed considerable exploitation of the 'green' label and significant rhetoric in firms' claims of being responsible citizens. Although all the companies analysed expressed concern for the environment, the principle motivation for many was a reactive response to stakeholder and regulatory pressures and an opportunity to grow revenue, cut costs and improve their image. These types of findings are common in the CSR literature and indicates that in many firms, CSR is 'skin deep', compliance based and forcibly rather than voluntarily embedded into strategic thinking and corporate culture. This picture is further reinforced by Banerjee (2007) who found that many glossy CSR reports are a form of 'green washing' that obscure the realities of corporate strategy and action.

Werther and Chandler (2006) point out that the main business arguments against CSR revolve around the issue of competing claims over the role of profit, competitive disadvantage, competence, fairness and legitimacy. In short, this is based on Friedman's view that the best way for an organisation to satisfy its obligations to society is to satisfy its obligations to itself and its shareholders. Although, as we discussed earlier, this view has its share of critics, objectively speaking, it is not without practical justification. For example, there is evidence that foreign direct investment in the developing world is shown to be one of the best vehicles to drive economic development, and as a by-product of that, comes better institutional governance, law and order, less corruption, and technology transfer to the host country in order to entice further inflow of foreign capital. Another argument against CSR is that it forces up costs. In particular, when CSR is not regulated or legislated for, then these costs are variable across a sector and competitive disadvantage could result from those firms that invest in CSR. It has also been argued that firms are not competent to deal with social issues and that this is best left to public organisations where these skills traditionally exist. Expecting private firms to undertake this role is not only inefficient but results in sub-optimal outcomes for society. There is also the argument that government is paid through business taxes to look after these issues and that it is their responsibility to do so. More recently, Brown and Fraser (2006) argue that CSR is difficult to operationalise and that it fuels insatiable public expectations which cannot be met, widening the legitimacy gap and backfiring on business. They argue that not only is reliable data not available to measure the real benefits of CSR to business and communities but that the doctrine of CSR has the potential to do real harm to the community and to even undermine the market economy by leading to a long-term redistribution of power and wealth away from investors who provide the capital for business expansion. These problems are further exacerbated by the complex and confusing array of non-standardised tools available for firms to measure CSR performance. This renders it difficult to tell what the real and overall picture is. While measuring profits is a straightforward activity for firms, measuring environmental and social impacts is a different story entirely. Nor is there any explicit indication of the degree of trade-off between the tri-factors before a business can be categorised as successfully sustainable.

Finally, there is also the 'paradox' of corporate performance and CSR whereby if a firm is doing particularly well financially, its efforts at doing good may be perceived negatively (Barnett 2008). For example, while a charitable donation of $5 million from a small firm might be received favourably, the same donation from a large and highly profitable firm might generate cynicism. The ethical legitimacy of such arguments is discussed further in Chapter 4. Thus, although CSR is enticing as a concept, there is a danger that it may only cause confusion unless the precise return on investment can be measured with some level of uniformity.

The relationship between business and society

In order to understand contemporary drivers of CSR, it is useful to have at least a basic grasp of the long-standing debate about the role of business in society which originated in the writings of Adam Smith (1776), Engels (1844), Marx (1867) and Lenin (1917). Adam Smith is often cited as the father of modern capitalism and one of the basic tenets of Smith's work is that the invisible hand of the market is capable of shaping self-interested behaviour and that this, in turn, plays a central role in providing human benefits and social goods. Smith made his point clearly when he wrote, 'it is not from the benevolence of the butcher, the brewer, or the baker that we expect our dinner, but from their regard to their own self-interest' (Smith 1776: 26–27). Economists since then have more or less accepted that when self-interest diminishes conditions that generate public welfare, then the market is deemed to have failed and external regulation is required. Indeed, this has been argued in relation to the global financial crisis which many argue was precipitated by a lack of regulation that in turn encouraged corporate excess and a flagrant attitude towards managing risk (Macfarlane 2008; Louisot 2009). Smith's ideas are associated strongly with the birth of the industrial revolution which also brought appalling living conditions, unsafe and inhumane working environments for men, women and children as young as five years old. This was fertile ground for social reformers like Marx, Engels and Lenin to document and deplore the conditions of the English working class and the downsides of the market system. Their work represents the earliest accounts of workers' resistance to exploitation, of their growing consciousness and will to struggle, and of their capacity through organisation to hit back at the all-powerful employing class. Following the industrial revolution, the field of management developed into a profession first shaped by Frederick Taylor (1911), an engineer who like most of his contemporaries conceived organisations as clinical goal-seeking machines which could be efficiently controlled. To Taylor, managers were the supreme coordinating authority responsible for creating a well-run, predictable organisation in which consistent decisions are made on established facts and principles and in which people behave as required on the basis of performance-related rewards. As Sikula (1996) points out, there was no room for social responsibility in this system since it promoted 'the ends justifies the means', and ignored ethics and morality as factors and inputs into production. Importantly, this literature also established the link between worker satisfaction and productivity. It recognised that workers had a certain degree of autonomy and that an organisation was made up of varying interest groups with multiple goals that would not always coincide with the organisation's. This resulted in a very different portrait of an effective manager who was responsive to varying interests within and outside an organisation and who could develop a shared set of values to motivate people to make the organisation a success.

Following these general developments in management theory, scholars specifically interested in the field of business–society relations were beginning to emerge, particularly in the areas of political science, economics, law and business policy. The modern era of CSR scholarship can be traced back to this period and in particular the publication of Howard Bowen's *Social Responsibilities of the Businessman* in 1953 and the political awakening of the 1960s to the threats of world food shortages and issues such as global warming which started a comprehensive ecological debate and put new pressures on the corporate world to become more responsible (Saha and Darnton 2005; Hill and Cassill 2004; Whetton *et al.* 2006). Early work during this period looked at *why* CSR was important and focused on the practical benefits of good business–society relationships such as enhanced reputation among potential customers, limiting excessive government regulation, and attracting and retaining high quality employees, etc.

Unfortunately, the early CSR literature offered little advice about how organisations should strategically fulfil their social responsibilities. But by the 1970s other practical mechanisms started to emerge such as: establishing departments of public/social affairs for managing community relations; corporate philanthropy; identifying and forecasting social issues; issues management; social reporting; changes in organisational structure; policies and systems, etc. In the 1980s stakeholder theory also emerged as a useful concept to explore the relationships between firms and their stakeholders. In particular, Freeman (1984) argued that the free market does not address wealth distributional consequences for minorities whose individual rights tend to be sacrificed for the benefit of the majority. This work asserted broader corporate responsibilities to stakeholders, which included: employees, customers, communities, consumers, suppliers along with shareholders.

In parallel with the above, CSR scholars also began to explore the relationship between government and business, in particular: the regulatory role of government in controlling the actions of business; and the involvement of business to affect such actions. Research in this area addressed the relative effectiveness of regulatory and voluntary CSR regimes, corporate crime and the factors that induce firms to comply or not with corporate laws and regulations. It also focused on tactics used by companies to influence government policy such as lobbying, direct political contributions, formation of political coalitions and trade associations, and the political strategies of multinationals in developing countries, etc. Out of this debate has emerged the notion of social contracts to extend company responsibilities to groups which significantly affect, or are affected by, a company's activities. This was influenced by the idea that different organisations have very different levels of responsibility to society which is determined by a range of factors: *causation, capability, awareness* and *proximity* (Oury 2007). For example, if an organisation's activities could be a direct *cause* of harm then they should hold a higher level of responsibility than those which cannot cause direct harm. The *capability* principle implies that an organisation's capability and power to mitigate potential harm should be commensurate with their responsibility. The *awareness* principle links an organisation's knowledge of the potential harm to its level of responsibility and the *proximity* principle argues that those organisations physically closest to a society or situation should accept greater responsibility. For example, it could be argued that an Australian construction firm involved with an international project in China has considerable responsibility towards local residents who may be affected by environmental degradation because its activities have a direct causal influence, that they should have knowledge of the potential impacts and that they have the capability to control this through contractual and monitoring arrangements. But the company may counter-argue that they were not proximate to their operations and that local business partners, contractors and regulatory bodies were more proximate to the impacts and thus more responsible to control them. However, with globalisation and technology advances it is becoming evident that proximity is less of a defence against accusations of poor CSR than it may once have been. As recent prominent cases in the shoe apparel industries indicate (Nike for example), a company cannot exonerate itself from the actions and responsibilities of its supply chain partners, wherever they may be. Conversely, those in the supply chain cannot exonerate themselves from the unethical use of their products by their upstream customers.

Changing perceptions of CSR

From the above we can see that there is little novelty in the idea that a corporation has not only economic and legal obligations, but also certain responsibilities to society which extend beyond these obligations. What is interesting, however, is the fact that as a concept and subsequently as a business strategy, it has remained obscure until quite recently. This raises three questions: 1) does it mean that corporations have become inherently 'bad' or 'immoral'?; 2) what do corporations do now that is so different from, say, 50 years ago?; and 3) what has changed to make organisations take note of CSR as a component of business strategy?

In response to the first question, clearly it is unfair to suggest all construction and engineering firms disregard the social and ecological impact of their activities. Many companys maintain and promote consistently high standards of ethical behaviour and there is little doubt that the construction and engineering industries have a considerable positive influence on society by creating the infrastructure which underpins and enables our growing wealth and affluence. Many large construction and engineering companies are indeed investing many millions of dollars each year in protecting the environment, in community development programmes which drive social progress, laying new infrastructure and services, and helping third world economies by exposing them to new standards of work, education and health for local workers (Farah 2001). These activities were given international endorsement in 1999 when the United Nations introduced its Global Compact Initiative where hundreds of corporations from around the world signed up to advance and address issues such as human rights, the environment, anti-corruption regulations, etc. (Bies *et al.* 2008). Although there is some (albeit limited) representation from the construction and engineering sector, these types of initiatives are perceived by sceptics to be in response to criticisms and exposures of corporate malpractice and to abate government and community concerns, rather than a genuine concern for CSR. This raises the question of whether or not the public should be concerned with businesses' motives when they adopt CSR practices. Does it really matter in the big scheme of things, so long as there are net gains in social, cultural, and environmental goals? Similarly, does it matter to the company what their underlying motives are when they adopt CSR? These are pertinent questions which lie at the heart of the CSR debate and which differentiate the idealists who believe that it does, because it demonstrates moral intent and commitment, and the realists, on the other hand, who believe that it does not matter because responsible firms are required to develop a sound business case for any decision they make.

Boundaries of CSR

Regardless of intent, firms are ultimately judged on their CSR record by the perceived balance between profit levels and broader contributions to society. This relationship defines the boundaries within which the CSR performance of a company is often judged by society. For example, when profits are perceived to be excessive but a company cuts its workforce to reduce costs, it is not surprising that the public react badly and interpret this as serving the interests of wealthy shareholders at the expense of the normal people. While corporations have a legal responsibility to serve the interests of its shareholders, such seemingly insensitive behaviour raises in the public's mind, the legitimacy of the profit motive and of the appropriate levels of profit.

As in most industries, the majority of organisations in the construction and engineering industries are economic entities operating in a competitive environment which are devoted to the pursuit of profit in the most efficient manner. There is nothing unethical or defensible about this and it is not our intention here to question or try to make moral sense of the profit-motive of companies in the construction industry, nor any other companies for that matter. Unfortunately, all too often, the profit motive is used to judge a company's (lack of) moral integrity, encouraging the view among business people that the major motivation of those advocating CSR is one that is anti-business (Camenisch 1987; Werther and Chandler 2006; Kline 2005; Oury 2007). This tension is exacerbated by the fact that many of the solutions proposed by CSR proponents incur significant extra costs with no apparent return on investment. The question of whether or not firms, be it in the construction and engineering industries or elsewhere, should be driven by the profit motive is a misplaced one because they are indeed business organisations created to generate wealth – they are not social or not-for-profit organisations. However, it *is* legitimate to question the level of profit, the way in which a business achieves its ends, how it makes its money, how it sells its services and products, and what obligations they have to society. Such questions raise numerous new ethical dilemmas for construction companies in their decision-making and strategising, for which many are unprepared. We will discuss the reasons for this and their implications further in Chapter 4. As we have witnessed in recent years the merits of profit maximisation, which has been the entrenched mantra of much business education and practice have been questioned with increasing public empowerment in the developed world. The mounting scientific evidence indicating the negative impacts that unrestrained business and economic activities have on the world is now more accepted and in many significant ways, the CSR 'movement' is now becoming mainstream, as illustrated in Figure 1.5, so much so that one can easily spot the wide array of voluntary CSR-related activities that corporations embark on – from philanthropic donations to looking after the welfare of employees, from fighting poverty to reducing global warming and improving human rights. However, the challenge is to link CSR to business strategy and, as the Economist (2008a) argues, master 'the art of doing well by doing good'.

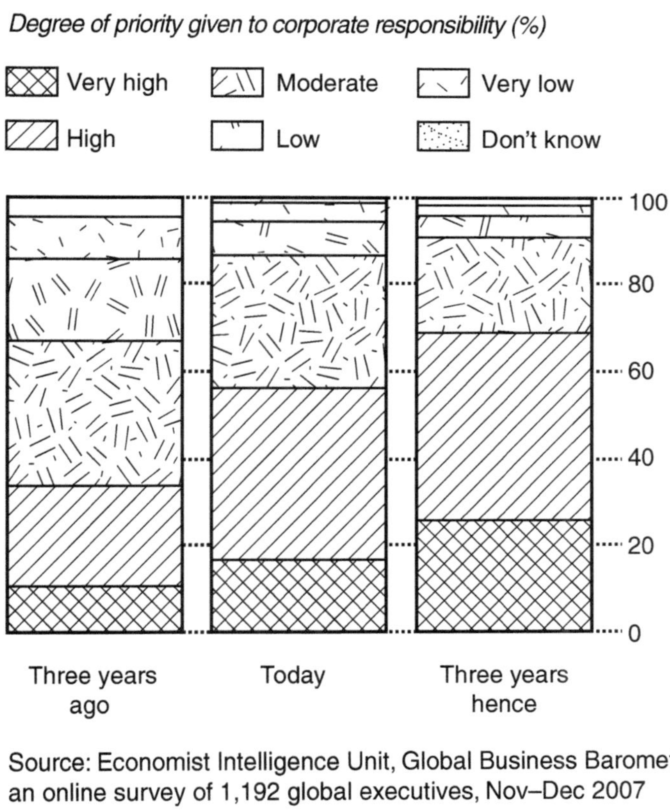

Figure 1.5 The growth of CSR as a priority

Source: Just Good Business, *The Economist*, 17 January 2008

The CSR record of the construction and engineering industries

In recent years, the focus on the construction industry has intensified. This is not surprising given the potential positive and negative role which the construction and engineering industries play in stimulating economic, social, cultural and ecological reform in developing and developed countries. On the positive side, at its best, the construction sector is a key driver of urban regeneration and economic prosperity through the creation of built environments that enable vibrant, wealthy and healthy communities to thrive. The construction sector is also a critical contributor to the economy, typically representing between 5–8 per cent of GDP and directly employing 8–10 per cent of the working population. The labour intensive nature of construction activity, its domestic structures in terms of labour and material supply and its widespread linkages to other industry sectors also ensures that investments in construction have a major multiplier effect to the rest of the economy (Lenzen 2001). In developing countries, where 95 per cent of urban growth will occur in the next two decades and where 80 per cent of the world's infrastructure will be built, the construction industry has a special role to play in promoting wealth and well-being and in achieving the UN Millennium Development Declaration goals to halve the number of people in poverty by 2015 goals (Ofori 2007; Zawdie and Murray 2009).

However, on the negative side, it has been estimated that the industry's products and processes consume up to 50 per cent of all resources taken from nature and contribute approximately 50 per cent of all waste going to landfills (Mak 1999). Cities house about half the world's populations but consume over three-quarters of the world's resources. UNEP (2009) estimates that over their lifespan, buildings are responsible for between 25–40 per cent of the world's energy use, 30–40 per cent of the world's solid waste generation, 30–40 per cent of the world's global greenhouse gas emissions, 33 per cent of the world's resources, and 8–12 per cent of water use. With carbon trading or tax schemes being developed in many countries, the need to include sustainability considerations in all aspects of the industry's activities will become both an economic and ecological imperative in the future.

In addition to the above, the construction industry has an unenviable reputation for protecting its workers from harm. For example, in Australia (which has a better safety record than many countries), about one person per week is killed on a construction project and the industry has a rate of serious injury which is 50 per cent above the all industry average. Furthermore, according to Lingard *et al.* (2009), the industry requires people to work harmfully long hours and provides fewer employment benefits such as holiday leave, sick leave and workers compensation than the all industry average. Despite the construction industry being one of Australia's largest and most culturally diverse (20 per cent of employees were born overseas (DIAC 2009)), evidence also exists that the industry is struggling to integrate different cultural groups into a harmonious and productive workplace. Recent research found that many employees experience workplace discrimination and that workers from ethnic minority backgrounds are exposed to significantly greater safety risks on site than other workers (Phua *et al.* 2010).

Globally, according to Transparency International, fraud and corruption would also appear to be an endemic problem in the industry. The Berlin-based anti-corruption watchdog's 2005 Global Corruption report labelled construction as consistently the most corrupt industry in the world (Transparency International 2005) and in 2008 the construction industry topped its Bribe Payers Index (www.transparency.org). According to the report, construction companies are the most corruption prone when dealing with the public sector and are the most likely to try and influence the policies of governments. The potential for corruption occurs at every phase of a construction project and in every single link in the contractual chain, posing huge obstacles to socially, culturally, economically and ecologically sustainable development. Some examples of corruption include bribes to officials to have environmental standards relaxed, disposal and dumping of toxic materials in unapproved areas, and coercion of officials into participating in illegal logging of protected forests and fisheries. Beyond the Transparency International report which focuses on large international projects, it is not difficult to find evidence in numerous countries to support claims of unethical practices in domestic construction industries. For example, Sweet (2009) revealed how a recent police investigation in New York uncovered substantial mafia involvement in the US construction industry with many companies either being controlled by the mob or having to pay a mob tax in return for protection or permission to operate. In the UK, a persistent problem has been the prominence of 'cowboy builders' (Holt and Edwards 2005) and a survey of corruption in the UK Construction Industry found that 41 per cent of those working within the industry considered corrupt practice to be widespread (CIOB 2006). More recently, following one of the largest ever Competition Act investigations in the UK, 112 leading construction firms were implicated in bid-rigging activities and cover-pricing.

Finally, the relatively recent development of Public Private Partnerships (PPPs) has highlighted the important social responsibility that the construction and engineering industries have to society. PPPs represent long-term arrangements to deliver services to society which have been traditionally delivered by the public sector. Through PPPs, the construction sector is more intimately involved than ever in the delivery of sensitive and critical public infrastructure and services such as education, health, water, electricity and security. Social infrastructure projects such as schools and hospitals arouse special community interest and are accompanied by an increasing public expectation that shared project ownership means shared project information, risks and rewards. Unfortunately, there is accumulating evidence that while PPPs deliver financial benefits to the private consortiums involved, there remains doubt and cynicism about whether they deliver longer-term benefits to society, particularly in areas such as health and education (Baron 2007; Black 2007; Smith 1999; Nishtar 2004). Humphries (2007) pointed to considerable scepticism in the community about the lack of transparency, complexity and secrecy in such deals and of the advantages of PPPs, citing a number of high profile debacles which have cost tax payers and communities dearly, including tunnels, rail links, hospitals and freeways. He argues that the public do not trust PPPs and that they suspect that the mantra of better value for money 'conceals a reality that they are being ripped off and that the big end of town nests are being feathered'. It is critical for the future of the PPP market and indeed the CSR reputation of the construction industry that these projects are not seen to serve only the multitude of financial, legal and management consultants involved, but also the interests of society that use public services on a day-to-day basis. As Sharpe (2004:8) noted, while PPP projects might offer creative solutions to public infrastructure needs, at the core is a complex web of relationships among bureaucrats, politicians, media, employees, general public (local, national and sometimes global), labour and special interest groups – 'Any PPP lives or dies on its reputation with these people'.

The preponderance of negative publicity about the construction and engineering industries makes depressing reading for an industry that does so much good and which is so central to the world's future economic, social, cultural and ecological prosperity. This is particularly true for developing countries where it is estimated that 95 per cent of the world's urban development will occur in the next two decades (World Urban Forum 2006). By implication, the industry has an enormous responsibility to ensure that the resources used in this urbanisation process have maximum impact on reducing inequality and poverty, and promoting better health, well-being and economic prosperity.

CSR in a global business environment

Without doubt, the most passionate and emotive dimension of the CSR debate stems from the globalisation of economic trade made possible by the WTO and other organisations including ASEAN, OECD and APEC. These bodies are internationalising the way business is conducted around the world but at the same time, according to opponents, are promoting and exploiting huge economic, social and ecological inequalities between developed and developing nations (Werther and Chandler 2006). The basic argument of activists is that globalisation has eliminated the balance between business and society envisaged by Adam Smith in the eighteenth century, where social brakes to self-interest were meant to be ensured by a close dependency between local communities, business activities and profitability. These brakes no longer exist in this new era of international business, transnational corporations (TNCs), global economic institutions, international law, mass media and international communications. Fundamental changes in the relationship between firms and local communities has transformed the CSR debate as is evidenced by the huge number of NGOs which have emerged to reconnect local communities with the many remote TNCs that increasingly affect their lives. Many argue that TNCs have emerged as the dominant economic forces in society, fundamentally changing historical relationships between corporations, societies and nation states. Indeed, today, 35 of the world's top 50 economic entities are corporations rather than countries, ensuring that business ethics and social responsibility take on further significance. For example, General Motors has greater annual sales than the GNPs of Denmark, Thailand, Turkey, South Africa or Saudi Arabia. Similarly, Wal-Mart's turnover is larger than that of Poland, Ukraine, Portugal or Greece. Large organisations routinely lobby governments around the world to influence policy in their favour, and Spar and La Mure (2003) argue that the power of TNCs has grown to the point that activist groups, when they are seeking to amend a country's laws often now target corporations rather than governments. To further enhance their power, TNCs are now demanding that their increased responsibilities to society should attract increased rights – the same rights as individuals – such as free speech, political rights, additional property rights and freedom in relation to free trade.

While UN conventions, OECD guidelines and National Codes of Practice, and international and domestic laws on human rights, equality, health, welfare and safety, and environmental degradation exist to monitor and control business activities in most developed countries, they are often missing in third world countries where many TNCs now operate. Governments have not yet been able to reach a consensus regarding standard expectations of corporate behaviour for TNCs which can be incorporated in international corporate law, and it therefore remains a fact that the power of international corporate law to control their behaviour is extremely limited and generally applies only through the intermediary of national legal authorities. Although the world's biggest economies have established various conventions to combat and punish corruption and bribery (for example, www.oecd.org/daf/nocurruption), truly global norms and laws regarding socially responsible, corporate citizenship and ethical business conduct do not yet exist to regulate TNCs and their sometimes deliberate exploitative activities. To some extent, the inadequacy of international law has led to the increased activities of many NGOs which are now playing a bigger role in highlighting unethical business practices and increasing public awareness about the responsibilities of TNCs. Avoiding the potential damage to reputation from being exposed by one of these NGOs has become one of the primary motives of firms adopting a compliance-based approach to CSR. In this context, proponents of CSR argue that legislation is only of limited value in bringing about socially responsible decision-making, and that there must be a social and psychological contract between business and society, underpinned by voluntary acceptance of ethical business values and of societal responsibility.

The relevance of CSR to SMEs in the construction and engineering industries

One question stemming from the above discussion is the relevance of CSR to the many highly geared small and medium sized enterprises (SMEs) which dominate the construction industry. The nature and scale of CSR issues are likely to be very different for SMEs compared to those facing TNCs. For example, in the first instance, they are less likely to be targeted by NGOs. Furthermore, SMEs as part of the supply chain of larger companies are also likely to be more insulated (at least at first) from regulatory changes, particularly in the international realm. Finally, complying with CSR may not only involve potentially high compliance costs but problems in attaining what may be unachievable targets without the assistance of the larger partners. On the other hand, Draper (2000) argued that because SME employee and community relationships are likely to be more intimate than in larger firms and because the impact of any SME activities is likely to be more direct on these communities, this makes SMEs a good environment for CSR to flourish. Nevertheless, while CSR should be as relevant, if not more, to SMEs as it is to large companies, for most SMEs the incentives, resources, expertise and longer-term outlook to engage with and understand these issues are often missing. Thus encouraging SMEs to take CSR seriously is fraught with practical challenges which are exacerbated by the lack of examples of good practice that such firms can draw from. Most, if not all, of the existing literature on CSR implementation is focused on big firms and can appear overwhelmingly demanding in terms of resources for small firms. Not surprisingly, CSR is therefore likely to be perceived as a 'luxury' that many SMEs can ill afford when their *raison d'être* is about achieving short-term economic gains.

While it is tempting to dismiss the above arguments as excuses to let SMEs get away with any CSR initiatives, there is undoubtedly some practical foundation to them. Therefore, the way forward is to first understand and accept that there are distinctive structural and contextual differences between SMEs and large firms, and that these dictate strongly what is feasible and viable. Second, is to justify a CSR business case that SMEs can relate to and which is relevant to their type of business. Although the starting principle for adopting CSR might be the same, the 'road map' to CSR implementation, so to speak, is likely to be significantly different for a SME compared with that of a TNC. In recognition of this, initiatives such as the UK's CommunityMark (www.communitymark.org.uk) are being established to encourage and measure the impact on and engagement of SMEs in their local community.

Despite the important differences between SMEs and large firms, CSR will only be taken seriously in any firm, regardless of size, if it can demonstrate a business case in terms of how it helps achieve their organisational objectives. Any company can adopt any number of CSR activities, at the very least, on a piecemeal level, from time to time. However, this is unlikely to produce a sustained return on investment unless firms are able to link them to core corporate objectives and competencies. Consequently many firms, although engaged in CSR from a reactive perspective, are still not well placed to understand whether or not there is really a worthwhile business case for CSR or how to implement it strategically if a business case exists.

Measuring CSR performance – ensuring transparency and accountability

Vilanova *et al.* (2008) argue that firms will only become truly strategic when CSR can be convincingly linked to business performance and when financial analysts are able to adopt widely accepted criteria to evaluate CSR performance. The difficulty in measuring CSR performance (in particular social performance), balancing this against economic goals and presenting all this in a way which satisfies the interests of many stakeholder groups, makes strategic CSR a challenging task. That said, companies are increasingly being required to report their CSR initiatives and transparency and accountability have become the buzz words of effective corporate governance. For example, in January 2005 legislation came into force in the UK called the *Operating and Financial Review* (OFR) which requires all public UK companies to report their social and environmental initiatives to shareholders. In Australia, the ASX Corporate Governance Council produced the *Principles of Good Governance and Best Practice Recommendations* (2003), strengthening the requirements of financial auditing and laying down ten core principles of good corporate governance and responsibility to be demonstrated in annual reports.

To assist in the achievement of these measurement and reporting requirements, numerous CSR indexes have emerged in recent years which can be used by investors and firms to measure and report performance. While useful, it is important to point out that these methods for measuring CSR do not guarantee transparency and accountability and that proper governance has a bigger role than just complying with indexes. Nor does it engender the cultural change which is needed to genuinely bring about a change in practices and behaviour among the business community. Finally, it should also be pointed out that the rigour and universal acceptability of measures and standards for non-financial performance are a long way behind those which have been developed for financial performance of companies (Horrigan 2005).

Regardless of the shortcomings of these indexes, they have undoubtedly played a crucial role in encouraging firms to engage with CSR by providing a standard benchmark against which they can compare their relative performance. The choice of indexes to choose from is quite bewildering and cannot possibly be covered here. Some of the most widely used include indexes like Reputex in Australia which collect and publish data on the CSR practices of Australia's top firms ranking them as –A to AAA. Reputex uses four main criteria: corporate governance (ethical practices, disclosure, risk management processes, etc.); employee management (occupational health and safety and training, etc.); environmental management (environmentally sound processes and practices, waste, sustainable products, emissions, etc.); and social impact (philanthropy, support for causes, etc.). In the UK, the FTSE4Good index measures the performance of companies that meet globally recognised corporate responsibility standards and have transparent reporting policies and practices in human rights, social and stakeholder management and environmental risk management throughout a company's supply chain. Similarly, in the USA the Dow Jones Index Global (DJSI World) has an in-built sustainability index to guide investors as does the European DJSI STOXX index. While these indexes and guidelines only apply to listed companies at the moment, it is likely that they will be increasingly used and adapted by financiers in assessing the CSR performance and reputational risks associated with non-listed companies in the future. To this end, the UN developed the United Nations Principles for Responsible Investment (PRI) which is jointly managed by the United Nations Environment Programme Finance Initiative and the Global Compact. The PRI aims to help institutional investors integrate considerations of environmental, social and governance issues into investment decision-making and ownership practices, and thereby improve long-term returns to shareholders. Over 3,000 companies across different sectors from over 100 countries are now members of the UN Global Compact, which was established in 2000 to bring corporations together to establish and report on responsibilities, based on 10 core values that span from human rights, labour employment, social and environmental protection to anti-corruption. In other attempts to ensure consistency, in 2002 many of the world's largest financial institutions signed up to the Equator Principles (www.equator-principles.com) which require them to ensure that the projects they finance are not only profitable but are procured in a socially sustainable manner and reflect sound environmental practices. The Global Reporting Initiative (GRI) was also formally established in 2002 as an independent body, collaborating with the UN and representatives from business and community groups around the world to develop and implement global sustainability reporting guidelines (www.globalreporting.org). The GRI is used by many firms in the construction industry and has the vision that reporting on economic, environmental and social performance by all organisations becomes as routine and comparable as financial reporting.

With measurement and reporting comes the desire for recognition through certification and ISO is the pre-eminent body which produces globally recognised management systems and procedures which are certified through periodic audits such as ISO 14000 and ISO 26000. It is not our intention to list the growing number of CSR-related certification frameworks but some prominent examples include SA8000 which focuses on basic workplace rights in accordance with thirteen human rights conventions (www.sa-intl.org) and AA1000 which focuses on stakeholder engagement and produces a series of indicators, guidelines, targets and reporting systems needed to achieve certification (www.accountability.org.uk). There are also those set up for more specific reasons such as the Ethical Trade Initiative (ETI) (www.

ethicaltrade.org)which through independent verification provides certification to assure companies are managing their outsourcing, business partners and supply chains effectively. Finally, there are those industry-specific certification schemes which incorporate an element of CSR in their charter. In the construction industry this would include the Forestry Stewardship Council (FSC) certification scheme to promote responsible forest management around the world.

There have also been significant moves towards equating standards of CSR in the construction and engineering industries. For example, in 2001 FIDIC issued a policy and set of guidelines on business integrity management in the engineering consulting industry, leading to certification under ISO 9000. Similarly, at the World Economic Forum in 2003, 19 leading international construction companies from 15 countries with combined revenues of over US$70 billion, formed the Engineering and Construction Task Force to tackle corruption, and by 2004 had developed and signed up to the 'Business principles for countering bribery in the engineering and construction industries'.

Despite the proliferation of reporting indexes, guidelines and standards, in many countries there is still no universal statutory reporting standard or agreement on how to report CSR activities, which means that companies are free to report what they like. Consequently, more often than not, this means that CSR reports are distorted in their favour by exaggerating positive information and playing down the negatives. For example, Saha and Darnton (2005) found that many firms' reports of environmental activities vary widely in scope and regularity and only reflect the concerns of a firm at the time, providing little assurance that a firm's performance will continue into the future. Also, many CSR reports are devised for marketing purposes and make extensive use of get-out clauses such as 'where practicable'. Furthermore, because firms can set their own objectives and targets, they can make just small incremental improvements each year and there is normally no way of verifying the reliability and validity of the data reported. Indeed, research indicates significant variation in CSR engagement, practices and reporting across the construction and engineering industry (Jones *et al.* 2006, Petrovic-Lazarevic 2008). Although many companies claim to care about CSR, verifiable evidence is often hidden in hard-to-reach places or presented in large complex documents which are difficult to understand. Furthemore, few companies include key external stakeholders in business decisions or board processes or can legitimately claim to integrate CSR across their business, making relatively limited use of key performance indicators and general benchmarking exercises to clarify, monitor and drive performance.

Conclusion

This chapter has introduced basic concepts of CSR, discussed the historical development of CSR and highlighted the moral, rational and economic arguments underpinning it. Increasingly, in response to the fundamental changes in the relationship between business and society, organisations are being pressured, both directly and indirectly, to engage in more transparent and socially, culturally and ecologically responsible business practices. Strategic CSR is about providing services and products in a way which addresses the dilemmas among the competing interests of stakeholders, through voluntary and sustainable business practices that place society, the environment and profits on a level playing field in a way which goes beyond what is mandated by law. The construction industry has significant economic power and responsibility to potentially affect positive and permanent change in the world. In construction, CSR is being slowly recognised as a business imperative, as client needs and investor preferences require more disclosure and socially responsible business practices. In response, many companies are beginning to realise that CSR is a business opportunity rather than a risk. Nevertheless, while there are pockets of promising CSR initiatives in the construction industry, it is far from being a mainstream business activity. The construction and engineering industries continue to have low representation in various initiatives and it ranks relatively lowly on CSR indexes, being implicated in corruption scandals and receiving bad publicity about their negative impacts on the environment, the safety of their workers and the interest of minority groups. The challenge for the construction industry is to make CSR a proactive way of doing business that could lead to positive advantage to core business strategy rather than a reactive compliance-based response promoted by a few passionate enthusiasts, as is currently the case in most companies. None of the above implies that companies should stop seeking profits but it does provoke important questions about the social responsibilities of modern construction and engineering corporations.

In developing a strategic approach to CSR it is certainly not possible to simply follow prescriptive solutions, since deciding which stakeholders and issues matter will vary widely from company to company and will also change over time. A strategic approach to CSR necessarily requires certain levels of sensitivity, responsiveness and flexibility to exist in a company's decision-making structure, as well as the establishment of sound governance structures, vision, mission, strategy and tactics to provide businesses with a sustainable source of competitive advantage in the long-term. We tackle each of these issues in turn in subsequent chapters of this book.

2 Strategic corporate social responsibility

In this chapter we discuss CSR's role in the development of effective strategy and we critique different approaches to achieve this. In particular we explore the merits and limitations of stakeholder management and, giving emphasis to the construction and engineering industries, we discuss practical considerations and steps that should be taken in developing and implementing an effective CSR strategy.

Introduction

For many managers of firms in the construction and engineering industries, the concept of strategy remains disconnected from reality. There are several reasons for this. First, most management texts on strategy do not consider project-based industries like construction. Second, there is a lack of empirical research on strategy in the construction and engineering sector (Langford and Male 2001; Cheah and Gavin 2004; Green *et al.* 2008). Third, despite many years of research on business and management strategy, there is still a debate about what it actually constitutes. As Hambrick and Fredrickson (2005:51) point out, 'strategy has become a catch-all term used to mean whatever one wants it to mean'. According to Hubbard (2004) and Hammonds (2008) this confusion arises from four main problems. First, during the 1970s and 1980s many managers tried implementing strategy with mixed success, seeing it as an artificial and difficult exercise. Second, Japan's ascendance during the 1980s focused managers' minds on higher achieving operational efficiency through techniques such as Total Quality Management, Business Process Re-engineering and Just-In-Time, etc. Third, during the 1980s there emerged a view that having a business strategy created organisational inflexibility in times of unprecedented change. Finally, many competing theories and research have emerged about strategy development in practice. In this chapter we seek to shed light on some of these issues and to provide a framework for a more strategic approach to CSR.

Models of strategy and their relevance to CSR

The foundational definition of strategy is sometimes credited to Chandler (1962) who described a strategy as simply a statement which originates at the top of an organisation, which is passed down, and describes the organisation's long-term goals. Ansoff (1965) who is often called the father of strategic 'planning' subsequently used a systems approach to conceptualise strategy as a high-level decision-making process designed to 'fit' a firm to its external environment. To Ansoff, strategic management was a top-down planning function which should be detached from the minutia of day-to-day tasks. This he differentiated from *tactical management* which was about middle managers developing detailed plans to achieve the direction, and *operational management* which was about supervisors undertaking day-to-day tasks to put those plans into action. These early theories aligned with the scientific model of management proposed by Taylor which regarded the organisation as a triangle where strategists occupy the top level with primarily downward flows of decisions, directions, instructions, plans and information. In this model, strategy implementation followed rational strategic analysis in a linear sequential fashion and all information was presumed to be available for top management about a relatively stable external environment. By understanding this environment, it was possible to develop long-range plans based on the extrapolation of long-term trends. These plans were best developed by specialist planning departments which would then assist the CEO with the detailed task of strategic planning.

While authors like Chandler and Ansoff were responsible for first conceptualising strategy, their ideas largely remained theoretical and insulated from practice. This was until Porter (1980; 1985; 1987) popularised strategy by incorporating them all into simple digestible frameworks that could be used by practicing managers to maintain a position of competitive advantage in a changing business environment. Porter is best known for his three most influential models: *the value chain*; *the five forces of industry*; and *the generic strategy matrix*. The value chain model helps strategists consider how the various parts of a supply chain add value to the end product or service in different ways and in different amounts, thus identifying weak points which need reform. It was argued that the majority of value is added by a minority of firms in a supply chain and that to achieve competitive advantage is where resources should be focused. Porter's five forces model identifies the five main forces which determine a firm's margins: *bargaining power of suppliers; bargaining power of buyers; threat of new entrants; threat of substitute products;* and *rivalry between existing competitors*. Competitive advantage and margins are maximised in markets where all of these forces are minimised. Finally, Porter's generic strategy matrix focuses on how firms differentiate themselves from their competitors. Put simply, firms can develop competitiveness by adopting four possible strategies: *cost leadership* (competing on price by being the lowest cost producer across a whole industry or sector)*; focused* *cost leadership* (competing on price by being the lowest cost producer in a niche market segment where it enjoys advantage over competitors); *broad differentiation* (offering a differentiated product or services across a whole industry or sector); and *focused differentiation* (offering a differentiated product or services in a niche market segment where it enjoys advantage over competitors). To Porter, whose work is still highly influential to management practice, the essence of strategy is about making choices and trade-offs and about choosing to be different and defining how firms will be different, rather than trying to be all things to all people (Hammonds 2008). A company without a strategy will try anything and attempt to beat its rivals at their own game, which inevitably forces customers to choose between competing firms on the basis of price. This results in sub-optimal solutions for all firms because it forces them into price wars and destructive competition which undermines an industry's competitiveness and margins. This is a recurring problem identified by many commentators in the construction and engineering industries and could imply an absence of strategy in firms generally. However, it could also imply few opportunities exist in the industry for strategic difference between firms, although this is unlikely in industries as diverse, large and competitive as construction and engineering. Rather, it is argued that the problem is more likely to be a lack of critical strategic knowledge to recognise and identify different value propositions in the supply chain which create a territory in which companies seek to be unique (Langford and Male 2001).

Porter argued that strategy must have continuity, although he also recognised that 'inflection points' happen when environmental circumstances change so much as to force a change in strategy. However, he argued that these points are actually quite rare in most companies and certainly not continuous as many writers portrayed during the 1980s when continuous improvement, revolution and change became the new management mantra. This raises the question (which we discussed in Chapter 1) of whether CSR is indeed one of the inflection points to which Porter refers which should force a change in business strategy within the construction industry? We will return to this question later.

While Porter's ideas dominated the 1980s and 1990s, towards the end of this decade it became clear that by exclusively focusing on external factors and strategic analysis, firms failed to consider the internal challenge of implementation (Hubbard 2004). To address this deficiency, recent research has identified a range of internal factors that are just as relevant to competitiveness as Porter's external factors, such as: *capacity to innovate, brand equity, reputation, workplace relations* and more recently *commitment to CSR* (Vilanova *et al.* 2008). While it can be argued that firms building on these capabilities are in fact pursuing a differentiation strategy, the reality is that many of these internal forces are not as easily measured using Porter's framework, which focuses primarily on tangible parameters. So, in response, alternative and complementary theories of strategy have evolved which are potentially more useful for developing an effective CSR strategy.

For example, the resource-based view (RBV) of strategy argues that a firm is made up of a unique pool of resources, namely: *physical* (land, premises, plant, etc.); *financial* (cash, capital, etc.); *human* (people, experience and expertise, etc.); and *organisational* (culture, reputation, relations, etc.) which are greater than the sum of the parts and which interact in unique ways in a web of internal and external relationships. Competitive advantage is gained from ensuring that a firm's internal resources are valuable, rare, unique and difficult to substitute. According to the RBV, a firm focuses on creating competitive advantage in the market by building unique internal capabilities and core competencies based on its unique resources (Prahalad and Hamel 1990). It achieves market orientation by continually adjusting these core competencies in response to intelligence about its market and customers' values and preferences (Narver and Slater 1990). These ideas were further extended by Senge (1990) who developed the concept of the learning organisation which was based on the idea that firms need to understand the systematic causes of problems rather than just react to them. This laid the foundations of 'knowledge management' and introduced concepts of 'intellectual capital' and 'social capital' as important organisational resources to be considered in developing a successful business strategy (Stewart 1997). The RBV was also extended by Hart (1995) in an attempt to promote the idea that CSR can provide sustained competitive advantage. Hart argued that in the future, firms will gain competitive advantage by recognising, managing and leveraging, better than their competitors, the emerging resource constraints imposed by the biophysical environment. This will provide the basis for strategic advantage in numerous ways such as in *product stewardship* and *sustainable development*, enabling firms to enhance reputation, reduce social and environmental impacts and reduce costs at the same time.

Strategy in a more socially conscious world

Collectively, the models of strategy discussed above, argue that a firm's performance is influenced by the environment, by its strategy and by other internal components such as resources, competencies and organisational structures. Their main utility is in offering a prescriptive and easy-to-understand framework that identifies the key strategic management processes that are crucial to a firm's performance. However, in recent years, as the field of strategic management has evolved, new ideas have emerged to question the adequacy of these models to address a more socially and environmentally conscious, dynamic and uncertain business environment (Rumelt *et al.* 1994). In particular, it has been argued that traditional models of strategy neglect the critical role of social engagement and multiparty interactions in the strategic decision-making process and their implications for firm performance (Farjoun 2002). More importantly, in response to an increasingly unpredictable business environment, it is argued that a shift in strategic thinking is now required which incorporates more dynamic and organic models of strategy (Barnett and Burgelman 1996).

Mintzberg (1994) was one of the first to question the adequacy of traditional models of strategy in addressing these new challenges. He questioned whether strategy should be viewed as a non-inclusive, imposed, top-down process and pointed to the limitations of traditional profit-maximising models. Mintzberg recognised that although firms have a fundamental, long-term profit maximising strategy, they also have to pursue other conflicting goals and objectives which may lead to trade-offs between long-term and short-term performance. He also argued that senior managers were not as rational and logical as suggested. Instead they were influenced heavily by political forces, were better connected to lower levels than had been assumed and were not as aware of their business environment as implied by traditional models of strategy. This supported Pascale's (1984) view that good strategy is often built from the bottom-up and that in many cases, lower level managers are more in touch with certain parts of the environment than senior managers. Effective strategy must therefore recognise that many parts of an organisation have connectedness with the environment, ideas to contribute and an important role to play in strategy formulation. According to Pascale and Mintzberg the best strategy comes from companies which are constantly responsive to their environment through the opportunistic synthesis of ideas from all organisational levels rather than from a detached or analytical top-down process. This is especially the case in developing an effective CSR strategy since fast changing generational differences in attitudes, education and management approaches may make senior managers less sensitive to changes in the CSR environment than lower level employees (Kotler and Lee 2005; Oury 2007).

Another contribution of Mintzberg's work was to highlight the important distinction between *realised* and *intended* strategy and raise the question of whether realised strategies must always be planned and intended, as is implied in traditional strategic planning literature. As always, the answer will vary from firm to firm and it is one of finding the appropriate balance. For example, Minztberg's (1994) simple model of strategy illustrated in Figure 2.1 argues that in reality, realised strategies arise from a combination of deliberate, emergent and unrealised strategies. This is an idea endorsed by Hubbard *et al.* (2002) who argue that one characteristic of the six most successful firms in Australia is that they all have both 'clear' and 'fuzzy' strategies. In other words, successful organisations have a plan but do not always stick *exactly* to what they said they were going to do because they are constantly in search of opportunities for growth and can respond quickly when unexpected opportunities arise. This has also been supported by research in the construction and engineering industries. For example, Hillebrandt and Cannon (1990) argued that because construction and engineering firms are labour intensive, strategic planning must have a large emergent dimension. Effective strategy therefore involves a constant opportunistic juggle to match human resources and management skills to a constantly changing array of geographically dispersed projects over time. More recently, Green *et al.*'s (2008) research found that a construction firm's strategy is more often emergent than pre-planned and shaped by unexpected opportunities and maverick behaviour, rather than in response to any formal mechanism. Green *et al.* also found that while boards of directors may occasionally intervene with planned strategies, there was little evidence to suggest that formal strategic planning techniques were used or that they had much impact on enacted strategy.

What is important from the above discussion is that there are two extreme schools of thought relating to strategy. At one extreme is the highly mechanistic and structured view, while on the other there is a highly unstructured perspective. Neither extreme makes much sense in today's business environment and, in reality, firms will find themselves positioned somewhere along this continuum depending on the set of prevailing contextual circumstances they face (both internally and externally). For example, it is reasonable to assume that companies adopting CSR would be better positioned towards the more unstructured and emergent end of the strategy continuum. However, the reality remains that firms that adopt CSR mostly do so without linking it to their overall corporate objectives. This is echoed by Porter who pointed out that, despite the surge of interest in CSR, in most cases it remains 'too unfocused, too shotgun, too many supporting someone's pet project with no real connection to the business' (quoted in The Economist 2008a). It is not surprising, therefore, that a recent survey in 2007 by the consulting firm McKinsey of CEOs participating in the UN Global Compact initiative found that a large gap exists between companies' CSR aspirations and their actions, indicating that despite the rhetoric, most companies have yet to effectively link CSR into business strategy (see Figure 2.2).

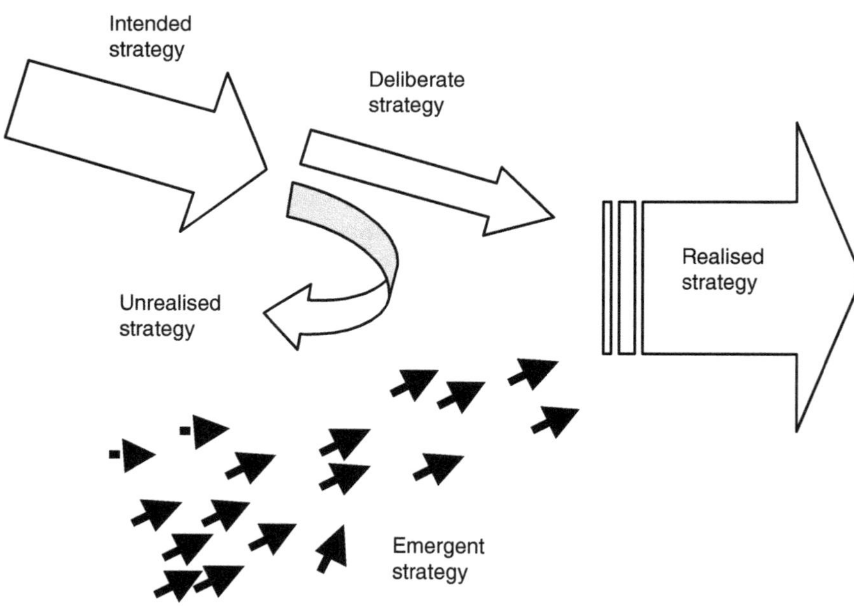

Figure 2.1 Elements of strategy

Source: Mintzberg 1994:24

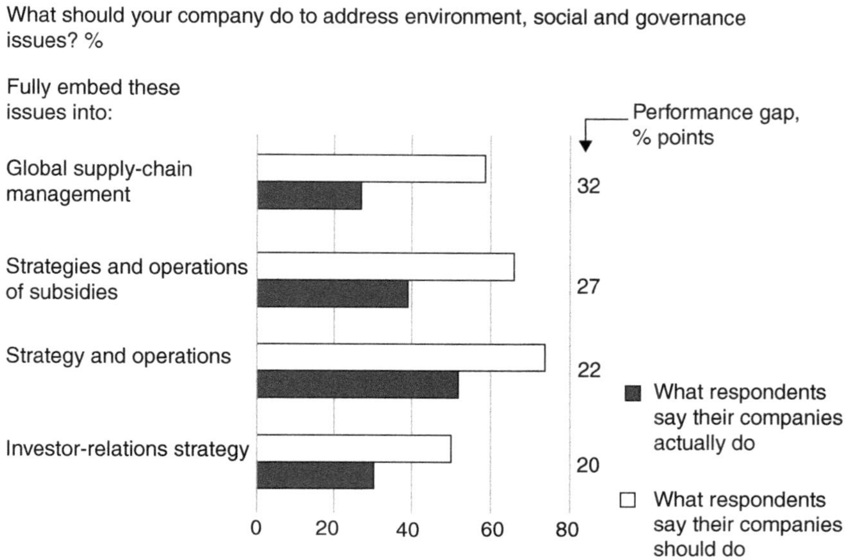

Figure 2.2 Gap between companies' CSR aspirations and their actions

Source: Just Good Business, *The Economist*, 17 January 2008

The point we are making here is that while businesses are now much more aware of the risks and opportunities that CSR presents, they are much less clear on what to do about them in a strategic sense. Until firms are able and willing to respond and act strategically to demands for greater CSR, and use it to add sustained value to their businesses, there will be continued scepticism in many quarters that CSR remains a largely public relations activity.

Incorporating stakeholder management into CSR strategy

In response to regulatory and community demands for greater CSR, more dynamic models of strategy have begun to emerge. One model which has its origins in the RBV of the firm, and which is gaining widespread attention is stakeholder management (Freeman 1984; Gao and Antolin 2004; Zsolnai 2006). The concept of stakeholder management was originally adopted in organisational studies by Freeman (1984) who defined a stakeholder as any person or group which can influence or be influenced by a firm, or which claim or have legal or moral rights and interests in the firm and its activities at any time in the past, present or future. This concept of stakeholder management goes far beyond simple models of employee participation and explores notions of social capital, arguing that firms can gain competitive advantage through developing unique, intangible social resources such as the relationships they have with customers, suppliers or employees, which are based on trust and reputation (its social capital). In this model, a firm is seen as a complex, dynamic and interdependent network of multidimensional relationships with a wide variety of stakeholders, the quality of which, can affect or be affected by its activities and vice versa. Performance and competiveness depend on how well firms manage and nurture these relationships strategically in order to achieve corporate objectives and how they are perceived to manage them by the stakeholders. Outside strategic objectives, firms also have a wider responsibility to society. These responsibilities are implied by government licences to operate and recognise that external stakeholders (other than shareholders) are also investors in a firm through its impact on their health, wealth and prosperity or by contributing intellectual, human or social capital (as employees for example).

The stakeholder management literature has shown how stakeholders with similar interests, claims or rights can be classified as belonging to certain categories of stakeholders to which a firm has different responsibilities. For example, stakeholders who can exert considerable influence on a firm, and are in turn most likely to be affected by its activities are commonly called 'primary stakeholders'. Without the participation of primary stakeholders, a firm will find it difficult to operate and in the long term, survive. On the other hand, 'secondary stakeholders' are not essential to a firm's operation and survival. Of course, primary and secondary stakeholders vary from firm to firm depending on their business objectives and also vary depending on the issue being addressed. But it is the identification of who the primary stakeholders are that is the key to strategic CSR, since excessive breadth in the number of stakeholders involved is neither practical nor sustainable. Hence, the two questions that must be asked in developing an effective CSR strategy are: who (or what) are the stakeholders of the firm? And, to whom (or what) do managers pay attention? In answering these questions, Mitchell *et al.* (1997:854) developed a detailed framework that defines the principle of who and what really counts. They propose that classes of stakeholders can be identified by their possession of three key attributes: 1) the stakeholder's *power* to influence the firm, 2) the *legitimacy* of the stakeholder's relationship with the firm, and 3) the *urgency* of the stakeholder's claim on the firm. In essence, a stakeholder's power is determined by the extent it can impose its will in a relationship, the extent to which its relationship with the firm is socially accepted and the extent that they can call for management's immediate attention. It is important to note that in the business environment, these variables are often in a dynamic state of flux. In reality, too, these attributes can be politically embedded, which often makes it difficult to untangle or decipher stakeholders into neat categories. Despite this, taken together, these three attributes provide a simple but practical identification typology that explains *stakeholder salience* from which stakeholder relationships should be structured and managed. Figure 2.3 illustrates the stakeholder typology in relation to the three key attributes. The higher the number attached to the identified stakeholder, the more salient it is (that is, firms should pay most attention to). For example, in the 'dormant stakeholder' category (1), although the stakeholders have power to impose their will on a firm, they do not having a legitimate relationship or an urgent claim and their power remains unused or symbolic. This contrasts with category 5, the 'dangerous stakeholders' who are characterised by having both power and urgency but who lack legitimacy. This group of stakeholders will use coercive means often illegitimately, to advance their claims, such as certain activist groups. In category 4, the 'dominant stakeholders' would typically be the shareholders or employees, while in category 7, the 'definitive stakeholders' are the people who possess all three key attributes of power, legitimacy and urgency, and top managers must give clear priority to them.

The above typology can inform CSR practice in important, practical ways. For example, by and large, firms tend to invest in more powerful stakeholder groups and to those offering most benefit to the bottom line. However, this may be detrimental to optimal CSR because when a firm focuses only on stakeholder power, it gives CSR the self-interested, amoral label. Equally, firms that focus only on issues of legitimacy move into the fuzzy morality of CSR. According to Mitchell *et al.* (1997:882), power, legitimacy and urgency must be attended to with measured considerations if managers are to serve the legal and moral interests of stakeholders.

While useful in helping managers decide which groups of stakeholders should be included in strategy development, Banerjee criticises the above framework for not helping to address inequalities in stakeholder engagement and for not explaining how and why some stakeholders are more powerful

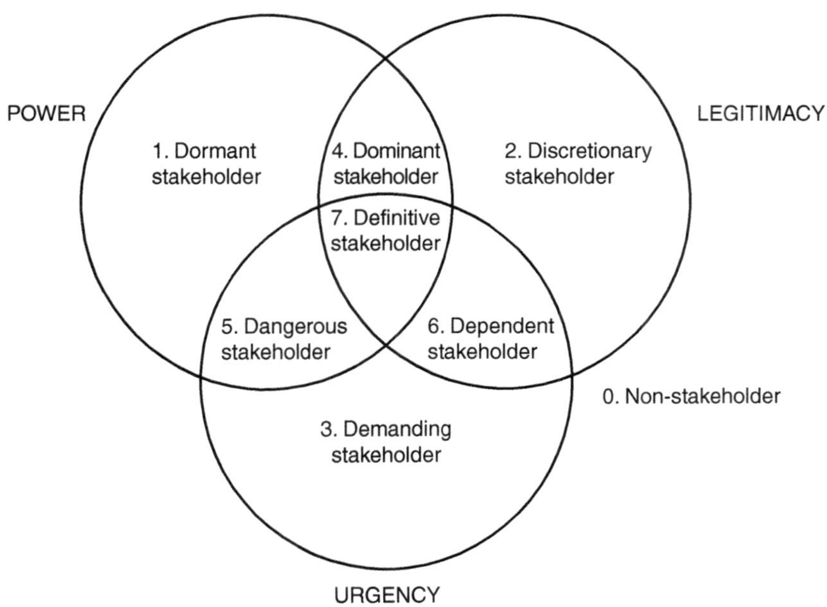

Figure 2.3 Stakeholder salience as a function of the three joint attributes of power, legitimacy and urgency

Source: Mitchell *et al.* 1997:874

or legitimate than others. As he points out, 'in a business firm, investors and shareholders are the only definitive stakeholders with high levels of power, legitimacy and urgency' (2007:33). Since to qualify as a definitive stakeholder one has to be in the possession of assets that are critical to a firm's success as well as power and influence to engage the firm, he argues that there are many disempowered groups (such as indigenous communities) who are adversely affected by corporate activities. Banerjee highlights an important potential conflict between strategy and equality because, to be strategic, firms have to be selective in terms of who they regard as salient stakeholders and if we accept that firms need to be strategic about managing stakeholder relationships, then we may have to accept that there will be potential inequalities. So, unless traditionally excluded minorities can acquire greater power, urgency and legitimacy, firms are unlikely to meaningfully include them in strategic decision-making. Teo (2009) describes how one group of Aboriginal activists attempted to do this by joining forces with local residents to fight the development of housing on culturally important coastal land which resulted in one of Australia's longest running, bitter, and as yet unresolved community protests against construction activity. While this group of activists have undoubtedly created inconvenience, delay, cost and damage to the reputation of the developer, they have constantly struggled to secure the controlling assets, power and influence needed to effectively determine strategic decisions about the development.

Working collaboratively with stakeholders

Another approach to measuring stakeholder salience considers whether or not a firm can *voluntarily* and effectively maintain *mutually beneficial* relationships with those parties (van Oosterhout *et al.* 2008:532). Based on this view, when a stakeholder can impose unilateral or mandatory obligations on the firm by using coercive force, then they should not be considered stakeholders of those firms. In a similar vein, no party should be granted stakeholder status that either lacks autonomy or ability to deliver on commitments. Most crucially, there should not be parties whose interests cannot be aligned with those of the firm. For example, a construction or engineering firm ought not to include a stakeholder if its interests are in preventing all construction or engineering activity without consideration or desire to come up with alternative solutions. While NGO's such as Greenpeace have been disempowered for many years by following such a zero-sum policy, in recent years they have deliberately sought to moderate their ideological stance and work with rather than against business in addressing their objectives. They have also pursued strategies such as purchasing shares and using their growing membership base, communication mechanisms and credibility to affect change. One example of how NGOs and construction firms can work effectively to benefit shareholders and stakeholders is the partnership that emerged between Multiplex and Greenpeace on the Wembley football stadium project. As Longshaw *et al.*

(2005) point out, Multiplex did not cynically engage with Greenpeace 'to provide air cover' but to develop a dialogue built on trust and genuine shared commitment to what they both considered the fundamentally important issue of illegally logged timber. From a CSR point of view, Multiplex's strategy of working in partnership with Greenpeace made good sense from a commercial, environmental and social perspective. Conversely, Greenpeace also achieved its objectives of avoiding the use of such timber.

Social capital and strategy

Cummings (2005) offers a postmodernist view of strategy by recognising that while one aspect of strategy must be focused on competitive advantage by reducing costs or increasing sales etc., it is also about recognising the value embedded in organisational networks, relationships, culture and company traditions and history. Recent trends and pressures to benchmark performance against one's competitors have caused many companies to underestimate the immense value of this social capital and the result has been an increasingly homogenous business world where most companies tend to offer the same products and services. In this homogenised business world, the corporate identity derived from the tacit knowledge contained within a company's social capital base can be more critical to competitive advantage. Without an identity and an ability to capitalise on one's own unique values, products, services and style, firms can only compete on price. Indeed, in the construction industry, because most firms' cost bases are broadly the same, competing on price really means competing on margin – a downward spiral which too often reduces performance to the lowest common denominator. This extends Porter's ideas of differentiation into a more culturally and socially grounded notion of 'identity'. It also extends Porter's idea of a supply chain into wider webs of relationships and social networks, a principle adopted by some firms in the construction industry such as Leighton's which have chosen to focus on relational contracting or an alliance approach to procurement, particularly on major infrastructure projects. The basic idea behind this approach is that a firm's unique connections in its supply chain and even with its direct competitors, creates a source of competitive advantage and that firms can mutually coexist by collaborating where they complement each other's competencies and skills. This avoids competing on price alone, improves margins and can also be viewed as a socially responsible way to do business (because there is an emphasis on openness, trust and collective responsibility). Indeed, even when analysed through Porter's generic five forces strategic framework, alliance contracting appears to be an effective way of increasing firms' competitive advantage because it reduces the bargaining power of suppliers and buyers, the threat of new entrants and substitute products and the extreme rivalry between competitors that has done so much damage to the construction and engineering industries.

Balancing CSR and profit in the construction sector

In the previous sections we have discussed the evolution of strategy and described the growing acceptance that intangible assets such as social capital tied up in stakeholder relationships are emerging as a key determinant of competitive advantage in an increasingly educated, empowered and informed, knowledge-based economy. As Venkatraman and Subramaniam (2006) point out, performance differentials between firms today appear to be more easily explained by intellectual and social assets rather than by monetary and physical assets as they were in the past (see Table 2.1).

While the traditional approach to strategy (often called strategic programming) appears less suited to the dynamics of the current business environment, it must be acknowledged that there are certain business environments in which this approach might be more effective. In particular, industries that 1) exist in

Table 2.1 The evolution of strategy

	Era 1 (circa 1970s)	*Era 2 (circa mid-1980s)*	*Era 3 (circa mid-1990s)*
Description	Portfolio of businesses	Portfolio of capabilities and resources	Portfolio of relationships
Key resources and drivers of competitive advantage	Economies of scale	Economies of scale and scope	Economies of scale, scope, knowledge and expertise
Key resources	Physical assets	Human capital and skills for managing relationships across businesses	Social capital as defined by position in a network
Unit of analysis	Business unit	Corporation	Networks of internal and external relationships
Key concept	Leverage industry imperfections	Leverage intangible resources	Leverage intellectual and social capital
Key questions	What products and markets?	What capabilities and resources?	What connections and knowledge?
Dominant view	Positioning	Inimitability of processes, resources and routines	Network centrality
Direction of strategy	Top down	Inside-out	Outside–in and bottom-up
Methods of strategy	Programming	Planning	Responding

Source: Adapted from Venkatraman and Subramaniam 2006

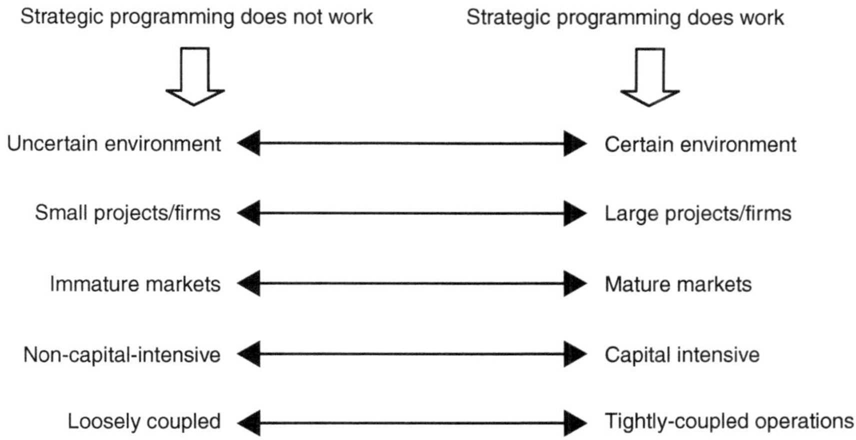

Figure 2.4 The effectiveness of strategic programming

relatively stable environments, 2) are mature, 3) have high capital intensity, 4) have closely knit structures, 5) are dominated by large firms and 6) have mainly mechanistic operations, tend to benefit more from a traditional strategic programming approach. However, given that the construction and engineering industries by and large do not possess these characteristics, one could therefore argue that the strategic programming approach which is adopted by many firms within the industry is inappropriate (see Figure 2.4).

Developing an effective CSR strategy

In devising an effective CSR strategy, managers of firms must ask themselves an array of difficult questions, the answers to which depend on a complex interplay of issues such as:

1 Why do good and what are the objectives? (for profit, for the sake of it, for a cause, to build a brand, to build a customer base, to build support for some future initiative, etc.)
2 What causes should be supported and why?
3 What stakeholder interests should be supported and why?
4 What resources are needed?
5 Is cultural change required? How to manage such change?
6 What are likely to be business partners' response and behaviour?
7 What initiatives should be developed that will benefit the chosen causes and the company?
8 How to integrate these initiatives into the business and corporate strategy?
9 How to implement a successful programme that will generate support internally and externally?
10 How and when to evaluate, measure success and return on investment?

These questions are discussed below but obviously the answers will likely be unique to every firm and ultimately depend on how senior management rationalise and justify their firm's core values and belief systems.

Practically speaking, it is the role of the chairman of the board to initiate the process of asking the above questions. Unlike in the past, when the time horizon for strategic planning was accepted to be about ten years, in today's business environment where issues and impacts are relatively urgent and short-term, it is now typically three to five years. The board of directors itself does not develop the strategy. This is done by the CEO and senior executives (managers) in consultation with primary stakeholders, with the board functioning as a governance, goal setting and monitoring mechanism and as a vital source of advice, experience, expertise and insight into broader external trends, risks and opportunities. It is the job of the CEO to ensure that these strategies are then implemented and are reported back to the board, which then reports to external stakeholders. Porter (1987) argued that the best CEOs are also good teachers of strategy. They make sure that everyone (employees, suppliers and customers) understand the strategy and the vision underpinning it. Effective CEOs are also able to implement strategic trade-offs between competing resources and goals. In developing a CSR strategy, it has been shown that these trade-offs are often difficult to achieve, making CEO leadership an especially important attribute.

In developing a CSR strategy there are numerous frameworks to follow. Most of those used in practice are based on traditional economics-based models of strategic planning which do not adequately suit the contemporary need for CSR. This is because they do not consider changing socio-political trends and expectations on construction and engineering firms and the need for greater resilience in an increasingly unpredictable business environment. To address these new challenges a practical approach to strategic planning is proposed which is more able to accommodate the development of an effective CSR strategy in the construction and engineering industries. This approach involves the following key considerations which are discussed in more detail below:

1 Define the organisation's CSR boundaries
2 Understanding the industry and broader business environment
3 Understanding the organisation
4 Understanding the stakeholders
5 Setting a feasible and appropriate CSR mission and vision
6 Developing a CSR culture
7 Setting CSR strategic goals
8 Developing a CSR strategic plan
9 Implementing strategy
10 Measuring strategic performance and CSR success
11 Communicating with stakeholders about success or failure.

Defining the organisation's CSR boundaries

Drawing a clear boundary around what CSR means to an organisation is the first step in developing any CSR strategy. In doing so, an organisation must define what it believes its responsibilities are to society in each of the categories identified in Figure 1.1. These will clearly be different for every organisation, but by defining and clarifying these responsibilities, a firm can better understand where the boundaries between its economic, legal, ethical and discretionary responsibilities lie. At the very least, having made these conscious decisions, they can then be evaluated objectively by the appropriate performance criteria. However, in a practical sense, this is not always easy and the point at which issues become the discretionary responsibility of organisations can be contentious. Such delineations are often blurred by society's perceptions of responsibility and the adequacy of laws and regulations which tend to change continuously. And there are also ethical dilemmas that are involved in making such judgements, which should be weighed against practical considerations in implementing them. Finally, as Clarkson (1995) argues, such choices are also influenced by market forces, which means that a firm's approach to CSR will be influenced most heavily by customers and clients who have the greatest and most direct impact on the bottom line.

Understanding the industry and broader business environment

Bettis (1998) defines an industry as a '*group of organisations or business units producing close substitutes*'. From a CSR viewpoint, it is very important to understand the forces in an industry driving and constraining CSR, how these forces are changing and expected to change over time and how they will affect future profitability. Many of the trends and forces driving CSR were discussed in Chapter 1 which will place a certain value on CSR initiatives in the future. However, the demand and supply for CSR and the types of CSR initiatives which are feasible are not only determined by the marketplace but also by the institutional constraints imposed by the government, society, profession, and industry. Over time, these forces tend to lead to industry homogeneity where firms in the same industry, facing a similar set of environmental conditions are likely to resemble each other, in terms of what they do and how they do it, in order to gain social and institutional legitimacy. So, at some point, a threshold will be reached where the adoption of innovative initiatives like CSR only provides legitimacy to the firm rather than improves its performance and competitive advantage. In other words, the strategic advantage to be gained from CSR is subject to a law of diminishing returns, as with every innovation, and firms that take the lead will benefit most. In strategic parlance, this is known as institutional or industry isomorphism (Dimaggio and Powell 1983) and the evidence presented in Chapter 1 would suggest that as far as the construction industry and CSR are concerned, this phenomenon seems a long way away. While there are some normative and coercive pressures from governments and societies to encourage firms to adopt certain CSR initiatives, the industry is still primarily driven by compliance and CSR is far from becoming the industry norm from which isomorphism can be said to have occurred. There are many reasons for this and until at least some are addressed, it is unlikely that isomorphism will occur in the construction and engineering industries, which means that there is still considerable potential for firms to derive competitive advantage from CSR. Some possible reasons include:

1 Clients and customers which do see the value in CSR.
2 The fragmentation of the industry and domination of SMEs which produce a short-term mindset.
3 The heavy reliance on short-term contract or casual employment practices which reduces responsibility for workers.
4 The project-based nature of the industry which creates a transient culture.
5 A confrontational industrial relations climate which reflects increasing tensions between competitive pressures and the rights of workers.
6 Extreme competitive pressures which produce a cost-driven rather than value-driven business environment.
7 The wide array of stakeholders that need to be considered over a project's life which make stable relationships difficult to nurture.
8 Procurement approaches which separate those who manage construction from the people who undertake it and are affected by it.
9 The flexible structuring of firms to enable expansion and contraction in response to cyclical supply and demand.

Understanding the organisation

In addition to understanding the exogenous factors that may drive CSR strategy, understanding internal organisational resources and core capabilities (endogenous factors) is also critical. These are a critical source of competitive advantage, especially those which cannot be easily imitated or substituted by competitors' and it is important to identify and capitalise on them in developing an effective CSR strategy.

Core capabilities

While resources refer to the tangible and intangible assets of an organisation (people, cash, land, buildings, equipment, reputation, IP, knowledge, social capital, etc.), capabilities refer to firms' ability to put those resources to good use, to achieve competitive advantage. This is dependent on *people's* individual skills, knowledge and abilities to do so, from the *policies and organisational values and norms* which direct their actions and determine their priorities and, from the *systems* in place to bring those resources together in a coordinated way. Although accountants have traditionally tended to focus on tangible assets when valuing a business, Chapter 1 argued that there is now an increasing realisation that intangible CSR-related assets are of equal value to competitive advantage. In construction and engineering firms, this was acknowledged by Langford and Male (2001) when they found that competitive advantage is especially dependent upon three core capabilities, namely a firm's: *relational assets, reputation* and *ability to innovate*. The challenge in building a CSR strategy in the construction and engineering industries is overcoming the likelihood that many of the required resources and capabilities do not readily exist, for the reasons discussed above. Thus developing the necessary CSR capabilities, resources and knowledge within an organisation is a critical aspect of the CSR strategy process in the construction and engineering industries.

Dynamic capabilities

Throughout this book, it has been argued that to be successful in the present business environment, firms need to be responsive and flexible enough to change and adapt to prevailing circumstances. This approach is supported by Cheah and Gavin (2004) whose empirical study of firm strategy and performance in the construction sector concluded that there is no universal formula for excellence. This emphasises the importance of 'dynamic capabilities', which refers to the particular (non-imitable and unique) capacity firms possess to respond most efficiently to changing environmental circumstances, be it markets, technologies, labour etc. Put simply, dynamic capabilities relate to a firm's ability to adapt in response to changing business environments in order to generate and exploit its unique internal and external firm-specific resources and core competences (Teece *et al.* 1997). This in turn depends on how a firm's internal managerial and organisational processes enable it to adapt its resources and capabilities to changing environments. According to Teece *et al.*, the ability to change these processes is constrained by the tradition and history of a firm's culture, policies and resources (its path dependencies). So for example, older construction firms with long established traditions and cultures (stronger path dependencies) may find it more difficult to adapt to CSR than younger firms. Green *et al.* (2008) argue that the existing literature in construction says very little about this and rarely are firms encouraged to think about how to adapt their organisational processes to changing circumstances such as CSR. Instead, according to Green *et al.*, construction firms are uncritically 'exhorted to adopt supposed universal best practice recipes such as lean thinking, partnering and integrated teams ...' and 'the overriding tendency is to conceptualise managerial skills as a static resource which is acquired and deployed as necessary' rather than developed and adapted to changing circumstances (p.65).

In terms of CSR, the scope and extent of what firms should do, must be considered in conjunction with the nature of their dynamic capabilities which need to be created, protected and continuously nurtured and improved to provide sustained competitive advantage. Hence, there is no hard-and-fast rule as to what should be the ideal type and level of CSR investments because it all boils down to the suite of core competencies, unique resources and dynamic capabilities that a particular firm has and the understanding of this will shape the strategic direction of the firm, including its CSR objectives.

Organisational maturity

Extending from the above, it is important for managers to understand that at different stages of organisational growth and development, firms will possess different core and dynamic capabilities. According to Greiner's (1972) seminal work, organisations move through a series of distinct stages of development, each of which contains a relatively calm period of growth (evolution) that ends with a management crisis (revolution). Because each stage of growth is influenced by the previous one, management, with a sense of its own organisational history and an understanding of its dynamic capabilities, can anticipate and prepare for the next developmental crisis and to turn it into opportunities for future growth. Figure 2.5 illustrates Greiner's model of organisational growth which shows that each evolutionary period is characterised by the dominant management style used to achieve growth, while each revolutionary period is characterised by the dominant management problem that must be solved before growth can continue. Greiner pointed out that the critical task for management in each revolutionary period is to find a new set of organisational practices that will become the basis for managing the next period of evolutionary growth. Eventually, though, these new practices sow their own seeds of decay and lead to another period of revolution. Companies in industries with moderate and high growth rates will experience these stages more rapidly than those in slower growing industries. The importance of this discussion is its suggestion that the timing of CSR strategy in any firm is likely to be determined by its unique stages of evolution and revolution, which in turn is determined by its growth rate, size, age and particular dynamic capabilities as well as various industry characteristics. Also, in developing a CSR strategy, this may be important because the jolt needed to convince a sceptical board or other primary stakeholders that CSR is important may indeed need to be a CSR crisis of some kind such as a scandal, fatality or a pollution incident that damages its reputation and bottom-line. Finally, Greiner's work suggests that the development of CSR strategy may move through

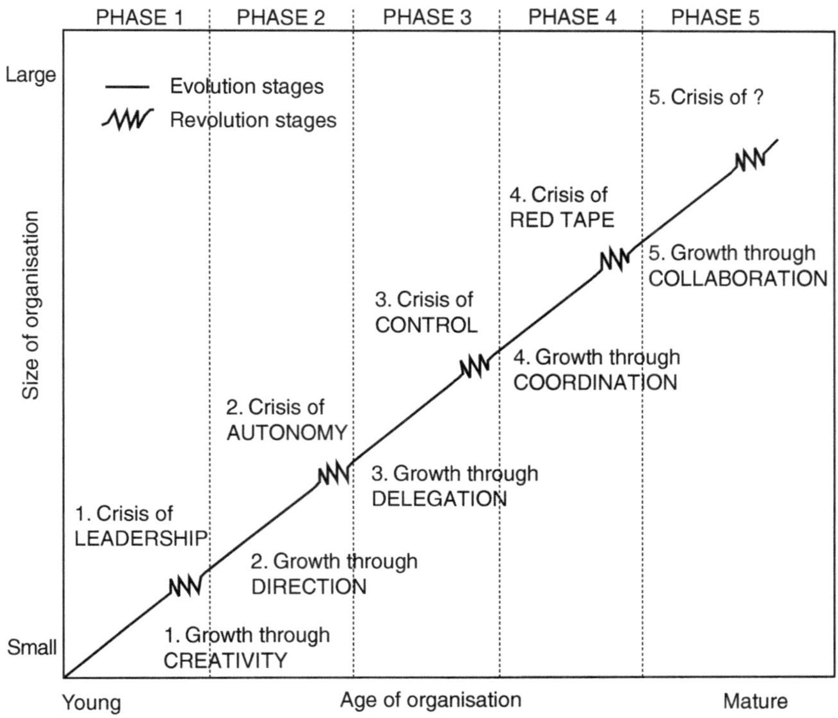

Figure 2.5 Model of the phases of organisational growth

Source: Greiner 1972

an initial growth phase of creativity into a crisis of leadership and then into growth through direction and another crisis of autonomy and so on. Understanding these dynamics enables CSR managers to anticipate strategies in advance to manage these phases smoothly.

Innovation strategy

Competitive advantage in CSR requires some degree of innovation. Some firms are followers and others are leaders and a firm will need to understand this in determining an appropriate approach to CSR. Useful in this regard is Miles and Snow's (1978) categorisation of organisations according to their strategy towards innovation which are briefly discussed below:

Reactors

Top managers frequently perceive change and uncertainty occurring in their organisational environment, but decide to 'wait and see' or are unable to respond effectively because their core and dynamic capabilities prevent it.

These organisations have little control over their external environment and tend to make adjustments only when forced to do so by environmental pressures. These organisations tend not to have formal, well-articulated CSR strategies. Even when management articulates a CSR strategy, resources, technology, structure and processes are unlikely to be supportive.

Defenders

These are organisations which have narrow product–market domains. Top managers in these type of organisations are highly experienced in their firm's limited area of activities, and they do not tend to search outside their narrow domains for new opportunities. These firms aggressively maintain prominence within chosen market segments and the main focus is on maintaining a secure niche in relatively stable markets. They also tend to ignore developments outside of the chosen domain but penetrate deeper into it, to exert dominance. In this type of firm, CSR would probably occur cautiously and incrementally.

Analysers

These are organisations that avoid excessive risks but excel in the delivery of new products or services. Typically they concentrate on a limited range of products and seek to outperform others on the basis of quality – the 'second but better' strategy. These types of firms operate routinely and efficiently through use of formalised structures and processes. They also monitor their competitors closely for new ideas and rapidly adopt those that appear to be most promising. Typically their strategy is to offer a mixture of products and markets, some stable, some not. In terms of CSR they are likely to be avid followers of change, quick to respond to competitors and often successful in imitating others through extensive market surveillance.

Prospectors

These are organisations that continually search for market opportunities and regularly experiment with emerging environmental trends. These organisations are often the creators of change and uncertainty to which their competitors must respond. Typically their strategy is broadly focused and in a continuous state of development. Growth stems primarily from new markets and new products and they tend to monitor a wide range of environmental conditions and trends, and be creators of change in their chosen industries. This type of firm would tend to lead CSR development in its industry, driven by the new market opportunities it provides.

Understanding the stakeholders

In Chapter 1 we discussed the potential benefits to be gained from engaging with stakeholders in the development of an effective CSR strategy. In essence, the likely benefit of effective stakeholder consultation is that more risks and opportunities are identified in creating a CSR strategy, more ideas for managing them effectively are generated and greater long-term support is secured for implementation. In contrast, to ignore stakeholders is to ignore that many have strategic power to influence the direction of a firm from a wide variety of sources and will probably attempt to do so anyway, incurring greater costs than if they were consulted in the first place. Furthermore, it would be contradictory not to consult stakeholders since the underlying purpose of CSR is to consider and manage the impact of a firm's activities on a wider range of stakeholders than only shareholders.

Unfortunately, we also provided evidence to indicate that many firms in construction and engineering still see investing in stakeholder relationships as risky and time consuming. Genuine stakeholder involvement that goes beyond perfunctory consultative processes is hard to achieve in reality because it requires management to be far more open to stakeholder influence than in the past. Furthermore, few firms have been able to appreciate and understand the complex stakeholder interactions that could shape their interests. This is due in part to the expanding range of stakeholders that appear to justify consulting in recent years and their growing ability to interact in new and complex ways to influence business outcomes. Even in a modest firm involved with modest projects, the network of stakeholders involved can be extensive and highly dynamic. However, stakeholder interaction is especially problematic on large construction and engineering projects where the ripple effect of development activity can extend globally.

Other likely barriers to effective stakeholder consultation include a lack of resources and time pressures on decision makers, which prevent them from engaging in meaningful consultation, and a lack of knowledge about how to do so effectively. Collectively, these factors create a significant gap between what theorists and educators claim is best practice and what actually occurs and can occur in reality in the construction and engineering industries. In order to overcome these barriers, Burby (2001) points to a series of critical questions which need to be asked in devising a practical and realistic stakeholder engagement strategy and these are discussed below:

- Which stakeholders have a legitimate right and power to influence strategy?
- What are our expectations and goals and what are theirs?
- What are their attitudes and values?
- When is the best time to involve them?
- What is the best way of involving them meaningfully?
- What information do we need to give them?
- Who to involve?

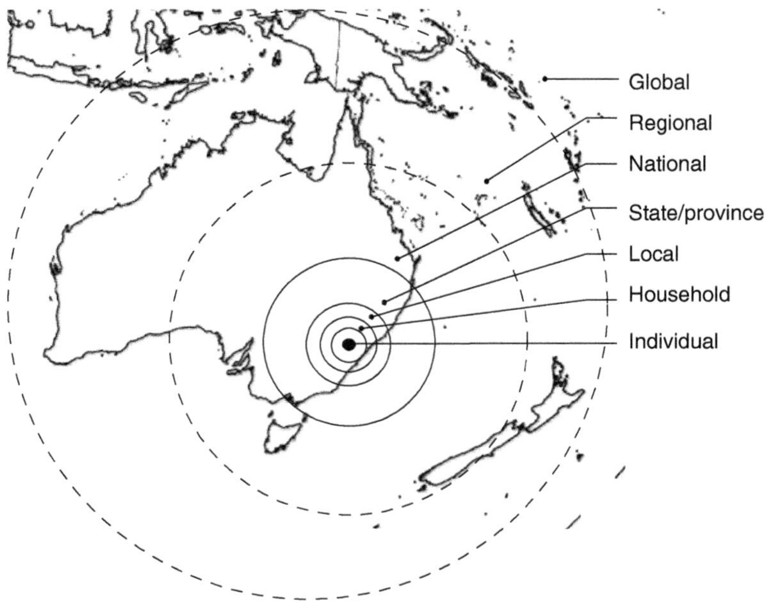

Figure 2.6 Potential ripple effects of development activity

Source: Kasperson *et al.* 2001:12

Clearly, despite claims of interest, legitimacy and representation from most stakeholders, every organisation has limited resources and time to achieve its goals. Thus it is not possible to consult every stakeholder all the time. Some are more important than others and need to be consulted in different ways, at different times and to different degrees. For these reasons, it is useful to adopt a stakeholder management strategy, which can disentangle the important stakeholders (in terms of their strategic importance to an organisation) from those which are less important. We have already pointed out that the framework provided in Figure 2.3 can be useful in doing this.

What are stakeholders' objectives?

Many stakeholder consultations are ineffective because the objectives driving the process are not clearly understood. Possible objectives for firms might include: compliance with regulatory requirements; giving stakeholders opportunities to voice opinions; educating and informing stakeholders about hazards and opportunities; educating and informing stakeholders about strategies to manage them; tapping stakeholder knowledge of hazards as a supplement to technical data; understanding stakeholder perceptions and preferences to deal with hazards; building a collaborative culture to mobilise a supportive constituency of stakeholders; and securing the participation and trust of stakeholders in the risk management process, etc.

Clearly, when the objective is simply compliance, engagement strategies are likely to be very different to when meaningful engagement is the aim. In contrast to compliance which may simply require a one-way media such as websites and newsletters, meaningful engagement will require more time-consuming two-way mechanisms such as community workshops and public hearings, etc. Burby (2001) found that the greatest dividends in stakeholder commitment to strategy resulted from pursuing a number of objectives, rather than just one.

When should they be involved?

Effective consultation requires effective planning and dedicated time. Ad hoc meetings attached to the end of other meetings send the wrong message to stakeholders, implying that the process is not meaningful and that their contributions are not valued. Clearly, decisions about timing are linked to the objectives underlying the consultation process. If the objective is simply compliance, then consultations will be infrequent whereas genuine consultation will involve far more frequent interaction throughout the strategic planning process. Burby (2001) found that the greatest dividends in commitment to strategy resulted from early stakeholder participation in decisions (up to 85 per cent higher than managers who initiated stakeholder consultation late in the process).

How to involve them?

Beder (1998) argues that Arnstein's (1969) ladder of citizen participation is a useful way of conceptualising the different ways in which communities can be involved in decisions involving controversial engineering projects. The ladder depicts a graded series of strategies with 'manipulation' being the least empowering form of consultation which essentially involves educating stakeholders to get them onside. At the next level of engagement, 'therapy' involves pretending to consult people to help them feel better about themselves. Next come Arnstein's 'tokenistic' methods of consultation which include 'informing', 'consultation' and 'placation'. Informing involves a one-way flow of information to stakeholders, consultation permits stakeholders to express their views but gives no guarantee they will be considered and placation involves token membership of boards and committees. The highest level of stakeholder participation in Arnstein's ladder are 'partnership', 'delegated power' and 'citizen control' which all involve genuine empowerment and redistribution of power to stakeholders. Partnership involves sharing risk and reward, delegated power involves handing partial authority for decisions to stakeholders and citizen control involves handing entire responsibility over. The most appropriate consultation technique in

Arnstein's ladder depends on a wide range of factors such as: time; cost and available resources; the maturity of stakeholders to handle responsibility; a firm's motives in consulting; available expertise and experience; the level of trust between stakeholders and the firm; and stakeholder expectations.

From a governance perspective, whatever approach is used, Matheson (2004) argues that it is useful to create a board strategy committee made up of an appropriate mix of the CEO, directors (executive and independent), senior management and other co-opted advisors and stakeholders. The task of this committee is to advise the board which is ultimately responsible and accountable for making strategic decisions, finalising the strategic plan and ensuring it is implemented by management through the oversight of the CEO.

What information to provide?

Access to adequate and appropriate information is essential in order to empower stakeholders, to secure their involvement in the strategic planning process and to ensure their acceptance of and commitment to any decisions made. Burby (2001) found that the more types of information provided by strategists, the more likely recommendations proposed would be accepted. However, there are a number of potential problems which can starve the stakeholder consultation process of essential information. As discussed above, some stakeholders may not be equipped to assimilate and understand the information being provided. In other situations there may be confidential issues which one party may not wish to divulge. The issue of information provision is therefore not just one of access and quantity but of content, context and trust. To overcome these problems, stakeholder relationships may need to be nurtured in numerous ways, stakeholders may have to be trained to interpret the information being provided and the information may need to be provided in a number of different forms and ways for it to be understood properly.

Setting a feasible and appropriate CSR mission and vision

The firm's CSR *mission* is the responsibility of the board and should also be clear, relevant and of practical value in providing direction to the directors and managers of the business. An effective mission should: be a common-sense of purpose and a 'rallying call' to the troops; be a statement of primary purpose; define 'why' the organisation exists, the reasons for its being, its core grassroot values and which stakeholders it serves; and give a sense of future direction.

The CSR *vision* is also the responsibility of the board but in contrast to the mission is a concise statement that articulates the 'ideal future' that the board wishes to bring about. The CSR vision should be inspiring and represent an easily visualised common image and shared vision of the organisation's future role and purpose in society and of 'what' it is aiming to be and to achieve. It should provide a sense of common direction for managers to develop and drive CSR strategy, and in this way should represent a critical link between the firms' aspirations, strategies and operational activities. Kotler and Lee (2005) argue that a lack of vision is the single most significant impediment to the growth of CSR in most firms and that many business leaders lack imagination and intrinsic motivation to incorporate it into their business strategy. According to Matheson (2004), this is due to a lack of independence on many boards which causes vision statements to be overly safe, politicised, uninspiring and tedious. This is likely to be a barrier to CSR in the construction sector since Chang *et al.* (2006) found that UK-listed construction companies demonstrated significantly lower levels of board independence than the top 50 listed companies in other industries. Indeed, Petrovic-Lazarevic's (2008) survey of CSR in many large construction and engineering firms would support this by highlighting most CSR strategies are simply restricted to the implementation of standards such as ISO 14001.

No matter how well crafted and collaboratively defined, the success of any vision ultimately depends on how well it is communicated and reinforced to ensure that employee and business partners can visualise and imagine the same future as CSR strategists. This should be done in as many different mediums and forums as possible, which may include formal meetings, training sessions, inductions for new employees and communications with existing employees, speeches by the CEO, company publications, annual reports, etc. Informal communications can be especially powerful in promoting and changing values so strategists should not restrict themselves to formal social networks.

Developing a CSR culture

The firm's mission and vision represent the building blocks of its culture which is an important enabler of strategy. Being subjective, difficult to measure, and difficult to communicate, the issue of organisational culture is often neglected in the strategic planning process. Yet it is widely accepted that to be effective, all organisations need to have clarity about what they stand for, and have a set of values that determine priorities and set the boundaries of accepted behaviour. This process starts with the CEO who should be selected by the board on the basis of alignment with the values which the board wishes to promote. Unlike organisational structure, practices and policies, culture is a personal, emotive and contagious thing which more than any other organisational attribute, is associated with the CEO's leadership, personal beliefs and behaviour. So it is critical that directors and the CEO are not only verbally supportive of the adopted culture but are seen to be endorsing it through their own behaviour, policies and strategies.

Setting CSR strategic goals

Ultimately, the board is held accountable for performance and thus having established a clear CSR mission and vision, the board needs to precisely identify, in measurable terms, the strategic goals it wants to achieve in order to realise it. These goals should be expressed in the form of key performance targets in the short-term (1–3 years), medium-term (3–5 years) and long-term (5–10 years). Setting clear goals which align with a firm's mission and vision statements is critical to ensure that the board can govern for performance which is its primary responsibility. They represent the standards against which CSR success will eventually be measured and will also represent the basis for the development of the detailed CSR strategies to achieve them.

As we have pointed out, traditionally strategic goals have focused on economic indicators of success but a CSR strategy will require them to also define and consider relationships with socio/cultural and ecological outcomes. In economic terms, goals may include areas such as profit, market share, sales, production levels, customer numbers, etc. In socio/cultural terms, goals might include occupational health and safety, community engagement, employee satisfaction, ethics, etc. In ecological terms, goals might include environmental preservation/degradation, waste minimisation, reuse/recycling, energy consumption, etc. It is also important to point out that these indicators need to be defined in *lead* and *lag* terms. Lead indicators represent goals that *enable* CSR strategy whereas lag indicators represent measures that CSR objectives have been achieved. For example, lead indicators may include the level of community involvement or buy-in in CSR initiatives, the number of staff participating in CSR programmes, the number of CSR managers trained and the number of CSR initiatives implemented, etc. In contrast, lag indicators will measure the end results of the initiatives such as water consumption, energy consumption, pollution incidents, emission levels, accidents, employee welfare, employee development and staff retention, etc.

The challenge in deciding on such goals is not only in ensuring that they align with the mission and vision statements but that they are realistic, achievable, sustainable and measurable. For economic goals, this is relatively easy given the well developed ratios, metrics and protocols that have been available over the years in the economics and accounting professions. However, this is not the case for socio/cultural and ecological outcomes where such experience and consensus do not yet exist. As discussed previously, many social outcomes are not easy to measure or indeed define, and there is also a temptation to set unrealistic goals and make promises which cannot be fulfilled. The difficulty and sometimes controversy in measuring many CSR goals makes it important, whenever possible, to refer to globally accepted charters and codes of practice such as the UN Global Compact discussed briefly in Chapter 1. This will ensure transparent and unbiased reporting, and convince stakeholders that what is being reported and measured has been externally verified and audited by independent reputable bodies.

Developing a CSR strategic plan

Statements of mission, vision and strategic goals describe where a firm wants to go but not the resources and specific initiatives which will be needed to get it there. This will require identifying appropriate programmes, milestones and timeframes to which budgets and financial structures must be linked. It will also require an understanding of the risks which may prevent effective implementation of the identified strategies, how they will be mitigated and identifying contingency plans to cope with their eventuality. Conversely, as is too often forgotten, it will also involve an understanding of the opportunities which might be grasped to enable the organisation to exceed its strategic goals.

Importantly, there are different levels of CSR strategy which need to be in alignment for all this to work, namely: *corporate, business* and *operational*. At a corporate level, *corporate CSR strategy* is about the future direction of the business which spans across an organisation's portfolio of business units. This level of strategy is concerned with how business units will grow or contract, the balance of resources between them and how the individual strategies of each business unit fit together into an integrated overall strategy. In a construction firm, separate business units may exist at a geographical level (regional or international), at a product level (infrastructure, housing, commercial, industrial, etc.) or at a functional level (marketing, HR, accounts, IT, development/finance, construction, R&D, property management and facilities management, etc.).

At the next level, *business CSR strategy* is concerned with how each individual business unit contributes to the overall corporate strategy in terms of products and services offered and the position of a firm relative to its competitors to create sustainable competitive advantage. For example, it may mean considering how the infrastructure division should compete in that sector, what position it should hold and when, where and how it should develop, etc. For the R&D department, for instance, it could revolve around the adoption of various technologies such as wind, solar or tidal to achieve sustainable construction. In IT, strategic decisions will likely involve how IT can be used as an enabler of CSR such as helping to improve CSR communications internally and externally. Finally, in HR, strategies will cover areas like recruitment, retention, training and performance management, etc. to support CSR objectives. Underlying these business strategies are investment decisions that involve budgeting, allocation of resources and the balancing of risk and return on investment to achieve corporate strategic CSR goals. However, as we show later in this book, the problem in making decisions about the type of business strategy to adopt is that many of the traditional mechanisms for investment analysis such as net present value (NPV) are not equipped to deal with the types of qualitative trade-offs necessary in making CSR-type decisions (Elkington 1999).

Finally there is *operational CSR strategy* which is mainly concerned with execution and implementation through the allocation and management of resources to achieve each business unit strategy. As Cheah and Gavin (2004) point out, strategy at this level addresses issues such as logistics, procurement, production processes and procedural functions for the provision of goods and services. This is also where the bulk of strategic research, knowledge and thus strategic management maturity has developed in the construction and engineering industries to deliver projects within time, cost and quality objectives.

CSR strategy at any level is underpinned by a combination of initiatives aimed to support different causes and in deciding what these should be Kotler and Lee (2005) offer some simple principles of good practice to follow:

Choose only a few causes to support

As we said earlier, the danger with CSR is that it can become a 'bottomless pit'. So it is essential up-front to define and limit the scope of initiatives and causes to pursue. Being selective will increase the chances of success by, for example, associating a company more strongly with an identifiable cause, allowing it to maximise the value-added of limited resources and to set clearer goals and outcomes. Being selective will also develop a strong constituency of community supporters and partners who can provide loyalty and benefits back to a business.

Choose causes which are of concern to primary stakeholders

Taking the time to identify causes which are of concern to primary stakeholders will enable firms to invest in constituents and communities which affect their own future in terms of potential revenues, human resources, customers, suppliers, etc. Furthermore, company employees and target audiences will be more likely to identify with the cause and work harder to serve it. This strategy is also more attractive to potential investors and will add credibility to statements in annual reports that a company understands the needs of important constituents and is giving back to them in some way.

Choose causes that align with mission, values and core business goals

Greatest success in CSR is likely to be achieved by employing initiatives which are a natural and logical extension of the business and company brand. Not only does this potentially maximise return on investment but it ensures that the CSR programme makes sense to important stakeholders who will be important to its success. They will therefore be less likely to be treated with scepticism and taken more seriously by investors. Employees will also be more able to understand the appropriateness of the CSR programme and be more motivated to engage with it.

Choose sustainable causes that can be supported and maintained in the long-term

Many social issues fall in and out of fashion over time. For example, during the 1980s addressing poverty in Africa was fashionable, in the 1990s the AIDS campaign was prominent and this decade seems to be emerging as one where sustainability is dominant. In response, the temptation is to switch causes to ride the wave of social concern and prominence. However, conviction, commitment and consistency are keys to a successful CSR programme since they avoid potential accusations of opportunism and maximise the chance of building strong and meaningful partnerships with community partners. Not only will this show a company is genuine but it will help to address the issues which are of concern. Many social issues are pervasive and it often takes a long time to make an impact.

Employ a range of integrated initiatives to support each cause

Most causes are best supported by a range of initiatives. However, it is important that the overall strategy should be coordinated and integrated. This not only increases the chances of success, but it brings otherwise disparate and competing business units and resources together in a positive way and gives a stronger message to potential benefactors, stakeholders and investors that the CSR strategy is thoroughly thought through and well managed.

Support a cause where there is previous experience

Supporting a cause where a company has a track-record and history allows it to build on the experience, lessons, social networks and reputation it has built up. It also allows it to capitalise on existing resources and knowledge such as people/partners with expertise in this area, existing marketing channels and technologies which may be brought to bear on the issue.

Match resources to return on investment

A strategic CSR programme is not a charity fund and if return on investment is not considered there is a good chance it will not be sustainable. To minimise financial risk, alternative funding options should be explored such as sharing costs with other competitors and business partners or agencies that have joint responsibilities to the same constituents. Finally, a company might establish corporate guidelines which may link resourcing levels to factors such as the number of employees, the nature of the cause, a certain percentage of pre-tax profits, a fixed dollar amount or even a sliding scale that varies depending on profit levels.

Implementing strategy

Implementation is often seen as the Achilles heel of strategy and many firms fail to achieve their intended strategic goals because they are not implemented properly (McIntosh and Thomas 2004). Common reasons for poor implementation of CSR initiatives include setting unrealistic timelines, not planning them effectively through clear actions, not identifying initiative owners and milestones, giving in to resistance, not adjusting plans in response to unexpected problems and not resourcing them appropriately. Although the board should charge the CEO with strategy implementation, this responsibility often rests with senior management, where the board is kept informed of its progress through the CEO. As we have discussed above, this requires that the organisation's strategy is clearly articulated, clearly defined in its objectives and intended outcomes, and clearly communicated to all primary stakeholders involved in its implementation. It will also need to be monitored and reviewed at regular intervals to ensure it remains aligned with changes in the business environment. Effective leadership and management are crucial and depending on the maturity of the firm, it is likely that the overall strategy will need to be phased or divided up into different parts relating to different divisions, tasks and individuals.

Measuring CSR success

Measuring performance and success is a critical yet poorly understood aspect of most CSR strategies. This is especially true when it comes to the social dimensions of performance which are far less developed, regulated and understood than the economic and environmental dimensions. As discussed above, CSR goals can be quite broad, difficult to define and quantify, success is often difficult to measure and benefits can often take some time to materialise. For example, how does a company measure the potential benefits to reputation of being associated with a particular cause? Furthermore, in most firms CSR success is in part defined by financial return on investment and one of the challenges of measuring this is the generally poor understanding of the cause and effect of many CSR initiatives on the bottom-line and difficulties in measuring in-kind costs such as staff time, providing advice, expertise, space, administrative services, dealing with enquiries, etc. There are many other challenges such as ensuring measures of performance are impartial and perceived to be so by different stakeholder groups. Because of these problems, in many companies CSR performance and return on investment are not measured at all, or done in an unsystematic and ad hoc manner, thus reinforcing further the notion that CSR can only be essentially a non-value adding, one-off philanthropic activity.

To address the above issues, provide legitimacy and to overcome problems of perceived impartiality many companies choose to employ a verifiable independent external agency to benchmark and measure CSR performance and return on investment. However, regardless of who does this, there are a number of key questions that need to be considered. The first question relates to the original strategic goals and KPIs which performance needs to be measured against. As we said earlier, in contrast to traditional models of strategy that tend to focus on financial and economic value-added indicators which look at short-term capital growth and dividend returns to shareholders, CSR strategies demand a broader set of social or environmental indicators. This has led some firms to use alternative indicators such as the balanced scorecard approach which can take into account, and hopefully balance short-term and long-term objectives, financial and non-financial measures, and lead and lag indicators from both internal and external perspectives (Kaplan and Norton 1992). The balanced scorecard concept is not new and is not a perfect tool for measuring CSR performance, but it does provide a broader range of perspectives than traditional accounting instruments, namely: *financial*, *customer*, *innovation and learning*, and *internal business*. Another popular approach to performance measurement that has arisen out of the CSR agenda in the late 1990s is triple bottom line, which seeks to balance measures of economic, social and environmental performance (Elkington 1999). Still in conceptual development, there is no agreement on what should be measured although as discussed previously, there are several organisations developing indicators to refine this approach such as the Global Reporting Initiative (GRI), World Business Council for Sustainable Development (WBCSD) and International Organisation for Standardisation (ISO). However, regardless of the approach adopted, the main unresolved question that lies at the heart of measuring CSR performance is the reliability, quality and timeliness of data. These parameters are well established in the realm of financial reporting and increasingly established in environmental reporting but far less so in social responsibility reporting. Inevitably, reliable data on CSR performance will take time to build up and when it does, the business case for strategic CSR will be far easier to make.

Communicating with stakeholders about success or failure

Developing a communications plan and conveying achievements or failures are steps that are often overlooked or done badly in CSR programmes. However, effective communications are essential in maximising the reputation and financial benefits of a CSR programme and will help to minimise resistance and secure continued buy-in to assist implementation. Openness and effective communication with stakeholders have become particularly important as the business world faces the prospect of a new raft of regulations and legislation governing disclosure, and new emissions trading and reporting schemes which will have major implications for business profitability throughout supply chains. As no one yet knows precisely what these new markets and regulations will mean for business, it is critical that businesses seek to understand and communicate the implications effectively to their clients, customers and business partners so that they can be prepared ahead of competitors.

Some important questions that need to be considered in developing an effective approach to communications will include: who is the target audience?; what are firms' objectives in communicating with them?; what mediums are best suited to this target group?; what timing is best and what style of communication will work best?; is it better to be brash and tell the world about one's achievements or is it better to be modest and let others do the talking?; etc. Each of these questions need careful consideration. For example, a company seen to be flaunting CSR initiatives can easily be perceived to be motivated by selfish goals whereas a company not promoting its work with different causes might be seen not to care.

Another important question that needs to be answered in developing an effective communication strategy is how to ensure the message 'sticks' with the intended audience and ideally spreads through its social networks to influence potential customers and other stakeholders. Perceptions of firms, both positive and negative, are contagious and can spread fast through social networks and once established are difficult to change. To ensure this happens in a positive way, some CSR-mature companies are involving influential primary stakeholders in designing CSR reports and in deciding what methods of communication are most effective to reach target groups. For example, Fleishman-Hillard (2009) found that online sources of information now outweigh traditional offline sources in obtaining information. Similarly, Edelman (2007) found that while newspapers are still the most trusted source of information (53 per cent), stakeholders are warming quickly to web-based media as a credible source of information (up from 29 per cent in 2006 to 34 per cent in 2007). Blog and social media usage is increasing rapidly but varies considerably around the world – from 48 per cent in China to 15 per cent in Australia – while radio and TV are the least trusted media largely because of widespread fear that the impartiality of reporting can be undermined in order to generate advertising revenue. Despite all of this, the most common way to formally report CSR activities remains through company annual reports. While it is not yet a legal requirement to report CSR initiatives in an annual report, many companies are voluntarily doing so. For example, according to a survey of the Global Fortune 250 companies, 80 per cent now publish CSR performance data in a standalone report which is up from 50 per cent in 2007 (KPMG 2008). However, this does vary greatly from country to country, where for example, the figure is only 20 per cent in Mexico but is 90 per cent in Japan.

While the KPMG report suggests that reporting on CSR is increasing, most is focused on environmental (not social) reporting and there is considerable confusion about how to report CSR effectively given the multitude of different approaches, frameworks and indexes which can be used to do so (Perrini *et al.* 2006). Most importantly, few firms have yet to fully establish the link between CSR goals and business performance, although many firms are now producing integrated annual reports with CSR sections to enable them to show these linkages more clearly.

Setting aside the various guidelines that are found in different reporting frameworks, the GRI guidelines are most widely used by the international business community, with more than three-quarters of the G250 and nearly seventy per cent of the G100 adopting them. The GRI provides detailed guidance on how to structure a report and readers are referred to (http://www.globalreporting.org/NR/rdonlyres/DDB9A2EA-7715-4E1A-9047-FD2FA8032762/0/G3_QuickReferenceSheet.pdf) for more information.Despite the existence of GRI and many other reporting guidelines, Fewings (2006) found a considerable lack of consistency in the way that sustainability initiatives were reported in the construction and engineering industries. Typically, companies ignore standard reporting models and decide themselves on what KPIs they will use to report performance. Fewings also found that construction companies tend to place greater emphasis on creating policies which might serve to boost their reputation (and presumably their bottom line) such as fighting corruption, community relationships and sustainability. Issues such as gender equality were typically given low priority, although equal opportunities and human rights featured more prominently in newer entrants to the sector, suggesting that this may become the trend in the future. Despite an apparent concern for ethical business practice, very few companies had a formal ethics policy and none were able to offer a definition of ethical behavior although some had developed policies relating to gifts, fraud, theft, etc.

Notwithstanding the trend for CSR reporting to become more standardised, at least in larger construction firms, concerns remain that CSR reporting may be open to manipulation and gives little confidence that a firm is behaving like a good corporate citizen (Basu and Palazzo 2008). A case in point is Enron, which appeared to be an exceptional corporate citizen, with many corporate social responsibility and business ethics tools and status symbols in place, but was far from it in practice (Sims and Brinkmann 2003). A problem in improving the state of CSR reporting in construction and engineering is that much of the guidance on CSR reporting, such as that produced by GRI, is so complex and demanding that it tends to be irrelevant to the vast majority of SMEs that dominate the industry. Given that much of what a large firm reports is likely to be related to the performance of its smaller business partners, reform currently falls on larger firms to help supply chain partners devise practical and realistic reporting systems which integrate with its wider responsibilities and reporting frameworks.

Conclusion

In this chapter we have sought to demonstrate that developing an effective CSR strategy is more than just a planning exercise and it involves firms making intentional, informed, collective and integrated choices about CSR initiatives that lead to positive performance consequences. More precisely, it is about adaptive coordination, which includes both responding to and influencing the CSR environment. We have argued that the stakeholder management approach is one of the cornerstones of CSR and that stakeholder management, as a strategy, is conducive to successful CSR. Increasingly, the way business is done today is determined more and more by the astute and careful management of a network of cooperative and competitive stakeholder interests which possess both tangible and intangible value to a firm. All things being equal, a company strategy built on its ability to accommodate, identify, evaluate and manage its diverse stakeholder interests is arguably more likely to result in positive outcomes. However, inertia to adopt this strategic framework can be expected due to several possible areas of doubts that firms have. First is the difficulty for firms to measure objectively the cause-effect relationship of stakeholder management and firm performance. Second, bounded by limitations of resources be it capital, intellectual, physical, human, social, or technical, the notion of investing/allocating resources up-front for some longer-term CSR benefits that are i) not easily quantifiable and, ii) not uniformly and objectively benchmarked, is often not the most appealing option. Third, stakeholder management when integrated properly to support CSR objectives requires a significant shift in the organisation's behaviour, values and culture systems. Fourth, challenges confront many firms in identifying, managing and nurturing stakeholder relationships that matter most, particularly for an industry like construction, where these relationships are many, transient and often highly political. Finally, for whatever practical or strategic reasons, when companies deliberately choose to adopt a more 'passive' approach to CSR such as philanthropic donations or taking part in ad-hoc community fundraising events, it is often very difficult for them to see the need to change radically the way they do things. These are indeed very challenging aspects facing every company regardless of size or type but large multinationals, with deeper pockets and a more systematic approach to see through such changes, tend to fare better than smaller firms which often lack the resources, commitment or conviction to see things through.

In including some examples, guidelines and frameworks to help firms overcome these challenges we are very much aware that these might be (wrongly) regarded as prescriptive, how-to lists. Given the diverse nature of firms, the industry and its corresponding environments, what we hope we have achieved for the reader in this chapter is a good balance between presenting the crucial discussions surrounding the main themes and providing some tangible indications about how these discussions translate into practice. To all intents and purposes, firms cannot be expected to transpose these guidelines without first thinking about and understanding their organisation, and consequently adapting them to suit what they do in their respective environments. As Pfeffer aptly suggested, 'What we do comes from what and how we think' (2005:128). Hopefully, this chapter has evoked that thinking process and we see the following chapters as a logical extension of this discussion where the focus will be on whether or not it is possible to combine governance structures and ethical concerns within a general CSR strategy for the construction industry.

3 Socially responsible corporate governance

In this chapter we argue that while company directors have a fiduciary responsibility to act in the interests of the corporation, the traditional interpretation that this only means shareholders' financial interests is potentially damaging to business interests. In doing this we critically review and appraise contemporary approaches to corporate governance from a CSR perspective, and advocate broadening its remit to include greater board independence and stakeholder involvement, underpinned by a longer-term, value-focused perspective.

Introduction

Systems of corporate governance have developed over time to provide shareholders with oversight of a company's operations. They provide the structure and mechanisms that officially manage the relationships between the shareholders, owners, managers, employees and other stakeholders (Micklethwait and Woolridge 2003). A key starting point to grasp in understanding corporate governance is that a company is a legal entity which exists independently of those who 'own' it. The board of directors is the central governance mechanism for this entity and its members are elected by the shareholders to direct the company's affairs in their interests. These directors are legally obliged in the first instance to work in the best interests of the company, not shareholders. To meet this 'fiduciary' responsibility, the board appoints the CEO and senior managers to manage the organisation on their behalf. Accountability to shareholders is then achieved through the company's demonstrated performance as reported through an organisation's formal reporting and communications structures, and through the monitoring and reporting of external auditors, government and regulatory agencies and NGOs. While in a practical sense the board does not manage the organisation, in most countries, company law vests ultimate responsibility for company management and performance with the board. Thus the capability of the board to direct the company, monitor performance, and direct and interact with the CEO and senior management is of crucial importance, as are internal governance structures and systems to monitor and report their performance. So any effective governance system should be effective at three levels. At the highest level there is the effective governance of boards by owners, external regulators and stakeholders, at the next level there is the effective governance of managers by boards and at the lowest level there is the effective governance of employees by managers.

Modern companies place enormous power in the hands of executives and while most act accountably and responsibly, some abuse this trust and succumb to the corrupting effect of this power (as exposed recently in scandals such as Lehman Brothers, etc.). It is to reduce this potential for acts of self-interest that systems of corporate governance have developed. Such systems are designed to install checks and balances on executive power and ensure it is placed in the hands of those who are best able to act in the interests of the company. However, one could easily argue, given events preceding the global financial crisis, that the system has glaring shortcomings that can be exploited by certain companies which are intent to do nothing more than employ a tick-box approach. However, on the other hand, there is also evidence to indicate that when robust corporate governance systems are in place and genuinely adhered to, they can enhance a firm's performance. For example, Matheson (2004) points to the reputation and direct financial premium associated with good governance in terms of higher financial performance. Bartholomeusz (2007) also pointed to the 'big' premiums that investors are willing to pay for firms who demonstrate good governance. Having said this, there is still an ongoing debate about the nature and extent to which corporate governance structures and corporate performance are directly linked. As vividly demonstrated in recent years by the collapse of many apparently healthy and highly reputable firms such as Enron, it is not simply a matter of getting the structural aspects of governance in place that is most important. Here, a complicit board was able to demonstrate compliance with formal regulatory corporate requirements and yet were fundamentally unsound in their governance procedures. As Horrigan (2005) points out, in reality the relationship between governance and performance depends much more on a complex interplay of many structural, behavioural, cultural, ethical, legal and discretionary factors. In this chapter we are cognisant of these relationships and seek to discuss them in more detail.

What is governance?

As Morck and Steier (2007) point out, governance is a variegated collection of different systems and approaches which range from the common 'shareholder primacy' model of governance in countries such as the USA, Australia, UK, Ireland, Germany, France and Finland, to the 'family model' of governance commonly found in countries such as Hong Kong, Mexico and Argentina. In the shareholder primacy model there is typically little concentrated ownership and most companies are controlled by thousands if not millions of middle-class shareholders who entrust governance to CEOs and other professional directors and managers. The quality of corporate governance is reflected in a company's share price and is monitored by governments, markets and regulators which try to shift the cost of monitoring away from shareholders by requiring firms to abide by complex laws governing disclosure, reporting, conflicts of interest and human rights, etc. Shareholders, governments and other legitimate stakeholders can sue, prosecute and even imprison directors if they break these rules. In contrast, in the family model of governance, firms are often controlled by a handful of wealthy families who may also control most of the country's corporations (sometimes its government as well). This is the most common system of corporate governance in the world and is often characterised by inadequate regulatory protection for stakeholders and pyramidal governance where the wealthy families exercise controlling rights over many layers of listed corporations (La Porta 1999). Of course, between these two extremes there are other governance models common to countries such as Sweden, Italy, New Zealand, Spain, Portugal, Denmark, Canada, Singapore and Belgium, where controlling interests are more balanced. Finally, there are countries such as Russia and China which are still in transition from centralised state controlled economies to partially decentralised market economies where models of governance are still evolving from centralised government control to public control. Morck and Steier (2007) argue that these global differences in corporate governance models have complex roots in different cultural, religious, governmental, legal and political traditions, geographies, financial histories, market developments and even accidents of history such as depressions, wars, colonisations, natural catastrophes and financial crises.

Hence, it is obvious that there is no widely accepted model of corporate governance and no one best way of governing for all organisations and countries. It is important to point out that this book takes a Western perspective and looks at the issue of governance from within a relatively well developed market economy. This is not to suggest that governance in this context is stable or superior since we have shown that construction and engineering corporations in the USA, UK and Australia, etc., and even corporations in other unrelated industries are experiencing great challenges in responding effectively to profound common changes related to governance issues such as CSR. While we acknowledge that all models of governance, whether they are formal or informal, or whether they are rooted in markets or rigid hierarchies, have certain common features that overlap with each other, it would be impractical to discuss each in detail here. So in this chapter we will be critiquing the shareholder primacy model and in doing so, drawing together the latest research and debates on how and why sound governance is important to CSR implementation in this context. Here we define corporate governance as the structures and systems that ensure there is sufficient oversight of boards of directors such that business obeys the rule of law and responds to the interests of customers, shareholders and wider stakeholders and society (Dimento and Geis 2005). Most fundamentally, governance is concerned with the stewardship of someone else's property (the owners'), the distribution of capital and other benefits within and between firms and society, and the responsibility of translating the shareholders', owners' and stakeholders' expectations and requirements into performance (Charan 2005). Hence, the two central features of governance are that the original *owners* of a firm give significant benefits and rights of control to *shareholders*, and that shareholders' rights and benefits are managed by a *board of directors* who have to balance their responsibilities to the firm, owner/shareholders, society and the environment. As Wearing (2005) points out, increasingly as we move into the twenty-first century, a distinct professional elite of directors are in control of our corporations and as these corporations grow by raising capital through the dilution of shareholdings even further, governance is increasingly critical to ensure they do the right thing.

Theories of corporate governance from a CSR perspective

Governance is an integral part of CSR because it provides the framework that can, in theory, protect stakeholders from the negative social, ecological and economic consequences of corporate misconduct. Two useful and competing theories used for better understanding the issue of corporate governance from a CSR perspective are *principal-agency theory* and *stakeholder theory*.

Principal-agency theory

According to principal-agency theory, the agency relationship is essentially a contract where the principals (owners/shareholders) employ and delegate one or more agents (directors/managers) to undertake a service on their behalf for some (usually financial) reward. Since agents are not likely to have the same interests as the principal, it is traditionally common for companies to offer them economic incentives in return for their performance – usually measured in terms of the company's profits and share price. Similarly, economic sanctions are usually put in place to deter managers and directors from behaving opportunistically. This 'carrot and stick' system of governance has by and large proved effective and has become the norm in many countries. However, principal-agency theory has also been criticised as inadequate by enabling the opportunistic behaviour of agents due to: 1) the owners in most cases (that is, remote shareholders) having incomplete or less information than the agents; and 2) the high costs associated with constantly monitoring the behaviour and actions of agents. Despite these criticisms, advocates of the shareholder primacy model argue that monitoring and control mechanisms do work and that agency theory can be effective in solving the problem of dispersed ownership by shareholders. It is argued that although each individual shareholder has a relatively small stake in the firm, the ease with which shareholders can sell and buy shares acts as a powerful deterrent to managers to engage in opportunistic behaviour because of potential vulnerability to takeovers, etc.

(Marris 1964). Nevertheless, critics argue that the principal-agency model of governance, which is based on neo-classical economic theories which see the corporation as a vehicle for shareholder wealth maximisation, is problematic. The recent financial crisis has added weight to these arguments in showing that managers do try to get around these rules to pursue their own self-interests rather than that of their owners'. It has also been shown that managers tend to focus too narrowly on the interests of shareholders to the exclusion of other non-profit making stakeholders who might have a legitimate interest in the firm's activities. For example, since share price is increasingly being used as a measure of executive performance, boards are often compelled to cut outgoings and expenditures, such as reducing their workforce just to get the share price up (Blair 1995). Indeed, in response to these criticisms, the Shareholder Association of Australia is currently encouraging boards to measure executive performance using short-term, medium-term *and* long-term indicators.

Stakeholder theory

As an alternative to the above, many scholars and regulators advocate stakeholder theory as a more effective approach to governance in today's business climate. This is because it encourages openness and transparency with a wider array of stakeholder interests and enables managers to focus on longer-term objectives which maximise value rather than just share price (Perrini *et al.* 2006). As Barnes (2002) and Galbraith (2006) point out, stakeholder theory is more in tune with contemporary social and regulatory realities in both domestic and global business environments which require far greater recognition of the value of human resources and business impacts on communities and the environment. As we discussed in Chapter 2, in relation to corporate strategy, stakeholder theory is based on the relational view of the firm and the idea of the 'extended enterprise' which requires firms to consider not only shareholders and directors (as explained in principal-agency theory) but other stakeholders such as employees, business partners, banks, insurers, suppliers, subcontractors, customers, governments and the wider community, etc. From a governance viewpoint, while principal-agency theory argues that it is primarily the shareholders who risk money on the firm and should receive controlling rights and reward in return if it performs well, stakeholder theory recognises that all stakeholders (to varying degrees) take risks when they work for a firm, or are directly and indirectly affected by its activities and are therefore entitled to some form of return also. It is argued that the intellectual, human and social capital they contribute is just as important to the firm as the financial capital that shareholders contribute (Lesser 2000).

Stakeholder theorists view corporations (especially large corporations) as quasi-public institutions with a licence to operate which is granted by society through their governmental regulatory and legal mechanisms. In response to being granted society's licence to operate, firms have responsibilities to the general public that go beyond their economic shareholders.

These include responsibilities to release information about their activities and to be evaluated in terms of both profit and other social objectives associated with a range of stakeholder interests, etc. The rationale is that stakeholders will reward or impose sanctions on organisations depending on whether accountability is perceived to be positive or not (Gray 2002). Sometimes, this is done through direct action and at other times through government intervention by regulating and monitoring the activities of firms. A good example of government asserting its role in businesses was the recent decision by G7 nations to bail out ailing banks in the current global financial crisis in exchange for better governance to reduce reckless corporate behaviour. Banks that have been 'rescued' by governments using tax payers' money have effectively been nationalised with various conditions attached, such as having the government's own appointee on boards and banning excessive executive pay and bonuses, although it appears that these criteria are far from being consistently reinforced or regulated.

The basis of the stakeholder theory argument is that 'corporate sustainability' (the capacity of the firm to continue operating over a long period of time) depends on its stakeholder relationships as well as its shareholder relationships. Boards should understand that there is a significant amount of 'social capital' tied up in these relationships which is of potentially immense value and which is yet to be exploited by most firms.

Alternative theories

While the ideals underlying stakeholder theory are fine, the reality can be quite different. While equity solutions such as profit-sharing schemes for internal stakeholders such as employees are quite well developed, less developed are those for other external stakeholders. As Wearing (2005) argues, stakeholder theory remains a theoretical construct in many ways and as yet provides no effective standard against which companies can be judged. It also provides no meaningful definition of what it means to 'balance' stakeholder interests with those of shareholders. According to Jensen (2001), stakeholder theory alone cannot make for better performing organisations because it lacks an economic logic – 'the theory asserts that firms should make decisions that take into account the interests of all the stakeholders in a firm, but without the objective function of increasing the long-term market value of the firm, the theory is incomplete' (p.236). For this reason, contemporary theories of corporate governance attempt to combine the best of both models to make them more practical. One such approach is referred to as 'enlightened value maximisation' which embraces the structure of stakeholder theory but accepts maximisation of the long-term value of the firm as the criterion for making the requisite trade-offs among its stakeholders. This approach identifies long-term value seeking as the firm's overarching objective (Jensen 2001).

Another balanced governance model is the 'enlightened shareholder approach' where companies adopt voluntary codes of conduct to manage their social and environmental impacts and governments adapt the law to ensure boards are held more accountable for their actions and wider business impacts (Eaglesham 2007). This approach is discussed in more detail later in this chapter under the section on corporate governance law. Nevertheless, it is relevant to say here that any approach that recognises the importance of stakeholder relations in organisations, demands that companies have appropriate governance systems that build and enhance these relationships, by measuring and assessing whether they are responding to their needs and to communicate this effectively to the outside world. These systems of governance need to broaden the traditional financial approaches to corporate performance measurement and reporting and need to recognise that if a set of stakeholder relations is important to strategic success then the creation of value cannot be limited to just one stakeholder group – the shareholders' bottom line (Boyce 2000).

In contrast to the balanced approaches discussed above, there are always those who argue for a much more 'radical' approach to governance. Advocates of this approach postulate a 'critical theory' that is highly sceptical of the potential for genuine corporate openness, transparency and accountability. They argue that there are inherent power differences and unequal resource distributions in society which will never enable traditionally disempowered stakeholders (such as indigenous and other minority groups) to effectively contribute to corporate decision-making (Lehman 2002). Critical theory is based on the belief that true dialogue, understanding and harmony only occurs between equals and current approaches to CSR are being manipulated by the all-powerful business community in a cynical attempt to cause confusion, promote their reputation and brand, and avoid regulation rather than to genuinely create change. Advocates of this approach argue that this problem is being exacerbated by enlightened stakeholder theorists who are prepared to engage with the business community in finding a 'middle ground' rather than engage in an ongoing social struggle to erode growing power imbalances in society. Supporters of this approach such as certain radical activists' groups, advocate violent protests, disruptive and adversarial campaigns using resources such as the internet to produce 'anti-accounting' or 'shadow-reporting' which highlight rather than reconcile business/society conflicts and expose the injustices of current systems of corporate governance and reporting. In contrast, critics of this approach argue that disruptive campaigns and 'militaristic' approaches do not bring about the desired outcomes of better governance or voluntary CSR and could backfire by enabling companies to legitimise their actions in excluding stakeholders from their decision-making (Driver and Thompson 2002).

The relevance of corporate governance to SMEs

The central role of shareholders and more recently stakeholders in the governance debate implies that governance is an issue which may appear of more relevance to large-sized public organisations than to the many SMEs and family-owned businesses that dominate the construction and engineering industries. Here there is often no apparent distinction between ownership and management. However, this misunderstands that the exact point at which ownership and control starts to separate (and thus the need for governance to emerge) differs from company to company and that size is not necessarily the deciding factor in the relevance of governance. The separation of ownership and control, and the accompanying need for effective governance is not a large-firm phenomenon and in many instances, sole entrepreneurs and small family businesses or individual entrepreneurs need to be just as concerned about it. As with large firms, owners of SMEs who rely on others to manage their business can also benefit from a well structured board to help advise and monitor them. It is good practice to ensure that this board has independent outside directors and a chairperson who can provide a range of competencies and experiences to help lead and develop a company's strategy and ask the difficult and important questions which managers might not ask of them. Increasingly, a majority of independent directors is considered a key feature of board performance and good governance, particularly in ensuring the success of CSR. This is because it increases the likelihood that the associated strategy will encompass a broader view of performance than simply economic goals. For example, Verrender (2009) when discussing the controversy of executive pay, points out that in many companies with weak chairpersons, together with a lack of independent directors and an absence of a dominant shareholder, it is possible for CEOs to control boards, stacking it and its subcommittees with supporters and forcing executive pay up without any scrutiny or control. For example, while in recent years shareholders have been permitted in Australia to vote on pay levels, the vote is non-binding which means directors can do what they want, often to the detriment of shareholder interests.

So it is clear that the structure of a board and the empowerment of shareholders, stakeholders, directors, and, in particular, independent directors is crucial to ensuring that firms act in the best interests of a wider range of stakeholders. It is also clear that the owners of any company, regardless of size, that have divested control to a board of directors, need an effective governance structure to ensure that their interests and those of legitimate stakeholders are being represented by the translation of their expectations and requirements into strategic plans, goals and ultimately performance.

The international relevance of corporate governance

Suggesting the relevance of corporate governance for companies, big and small, raises the question of whether the concept of trust – which is regarded and embraced by many to be the cornerstone of good business practice – still exists. Are we now so sceptical about businesses and are businesses so far removed from the communities in which they work, that we have to resort to complex systems of checks and balances to control their behaviour? For many small, family-owned businesses and in countries such as Japan, this seems like an odd idea when the success of these businesses rests on informality and stakeholder trust that have been built over generations. It is against the backdrop of these often complex but apparently successful stakeholder interactions that scholars have cautioned against a full scale export of the 'Western/Anglo-American' governance system in the name of globalisation particularly as the model fits poorly with many East Asian countries, and even some multinational firms (Kelly *et al.* 1997). A big distinction exists when reviewing the appropriateness of complex and formalised governance structures between UK or US firms and continental European or Japanese firms. For example, in the UK and USA, companies are largely characterised by dispersed ownership where the need for direct control could be argued to be greater than in continental European or Japanese firms where families typically retain a bigger capacity to exercise direct control and arguably have less need for externally imposed market-oriented rules and regulations (Aguilera and Jackson 2003). Nevertheless, while arguments of cultural relevance may have some merit, arguments about family or non-family ownership justifying different governance structures are highly debatable. This rests on the traditional notion that governance structures exist to protect shareholder interests whereas newer models of governance consider other stakeholder interests such as employees and the community. From this contemporary perspective, while a family firm may have fewer innocent shareholders to harm, through poor directorship and management, they still have the capacity to harm the community in as many ways as non-family firms, and thus justify governance controls.

Reframing corporate governance for CSR

As we have shown, the traditional shareholder-primacy model of corporate governance has been widely questioned, analysed and reframed by researchers, academics and policy makers in many disciplines in recent years. A highly diverse literature has evolved to enable us to probe the relationship between CSR and corporate governance in new ways, providing a number of basic continuums to reconcile the differing perspectives on philosophy and approach. Some of these are discussed below.

Profit and value continuum

Traditionally, models of corporate governance have been based on two fundamental assumptions. First, firms can maximise their contribution to society and be most efficient when managers seek to maximise residuals that the firm generates over time. Residuals are defined as the difference between what the firm pays for its inputs and what it gets paid for its outputs in a competitive market (taking into account the cost of externalities such as pollution, etc.). This residual (profit) influences share value and thus the ability of firms to raise capital by issuing new equity for growth. Second, firms (at least investor-owned firms) must distribute those residuals on a pro-rata basis to shareholders, again increasing the ability of firms to raise capital by issuing new equity (Fox and Heller 2006). In simple terms, defective governance in this traditional model means a firm not meeting one or both of these conditions. However, at the other end of the spectrum, critics of this approach and advocates of CSR emphasise the importance of maximising 'value' rather than simply 'profit'. To them, value is something which incorporates a broader range of objectives (economic, socio/cultural and ecological) which need to be balanced by effective governance. So in contrast to traditional models of governance where the board essentially acts in the financial interests of the owners and shareholders, the value-based approach implies a pluralist model of governance in which the actions of the board are driven by a wider set of interests.

Shareholder and community/stakeholder interests continuum

Another aspect of the governance debate questions the relative power differentials and distributional outcomes of business between shareholders and non-shareholders currently available through market mechanisms. Advocates of CSR argue that current governance mechanisms essentially distribute the majority of benefits to shareholders and provide non-shareholders with little power beyond their capacity to bargain and agitate. They also question traditional economic models of governance that frame shareholder and non-shareholder interests in oppositional terms and argue that in the absence of externalities, social welfare is maximised when each firm maximises its profit (Jensen 2001). In contrast, CSR advocates argue that a firm cannot serve society effectively or indeed maximise its profits if it ignores the interests of stakeholders. They propose an alternative model of corporate governance where the traditional role of the board changes from simply serving the financial interests of shareholders to also mediating the interdependent but often conflicting interests of stakeholders, all with legitimate but varying rights in influencing the direction of the company (Posner 1993; Deakin 2005).

Global and local standards continuum

As Charkham (1995) points out, corporate governance laws apply territorially, meaning companies and their international subsidiaries are subject to the laws of the land where they are located. Any company trading internationally with overseas subsidiaries will therefore need to understand the prevailing laws in the operating countries. However, internationally, corporate governance laws vary considerably in their coverage and rigour and a company also needs to understand the background to any laws and how they operate culturally. Many unscrupulous companies take advantage of poor corporate regulations in many parts of the developing world by shifting their operations there, without understanding the new cultural context in which they will have to operate and the perceptions of customers who buy their goods and services. This has led numerous multinational firms to not realise the economic benefits they initially sought and to suffer severe damage to their reputation from accusations of exploitation of natural and human resources by host countries. This in turn has led to calls for greater international controls over the activities of multinationals and fiercely contested debates about whether corporate governance models should reflect cultural and regional differences and traditions (Fox and Heller 2006).

Unitary and two-tier board continuum

At one end of the spectrum is the 'unitary' board structure which comprises a single board of external 'independent' directors and management representatives (executive directors). At the other end is the two-tier board structure under which management and supervisory roles are delegated to two separate boards: 1) a management board comprised entirely of management representatives with responsibility for operational issues; and 2) a supervisory board comprised of entirely external independent directors which has a higher level monitoring, advisory, policy and strategic role that sets the vision, values and direction for the firm. Advocates of CSR argue that a two-tier board structure is more conducive to responsible strategy since it allows improved oversight of the decision-making process and offers higher levels of true independence in corporate decisions.

Economic and psychological contract continuum

The economic perspective argues that a firm's activities consist of a set of potentially value-enhancing transactions between two or more parties (Fox and Heller 2006). It also argues that every transaction has associated costs (including governance and monitoring costs) which need to be minimised to maximise efficiency and profit. The in-built rewards and risks associated with each transaction are traditionally recorded and managed by legal contracts which stipulate the allocation of risks and rewards between the contractual parties.

However, advocates of CSR argue that this approach to governance has certain limitations in that formal contracts cannot define every aspect of business transactions, especially the less tangible socially constructed expectations and obligations that may be assumed to exist with some stakeholders. Instead, non-economic expectations are best managed using a *psychological contract* which reflects the beliefs and values of each party as to their mutual obligations within the contractual relationship (Loosemore *et al.* 2003). From a CSR perspective, the relevance of psychological contracts is in recognising that in addition to 'hard' deliverables, there are a 'soft' set of expectations which also have to be organised and managed. At a project level in the construction and engineering sector, we often refer to these as partnership arrangements. However, these are often narrowly framed in terms of business relationships whereas there should be similar arrangements with other legitimate stakeholders and communities.

Short and long-term ownership continuum

The role of ownership has changed in recent times to become more short-term in outlook. This is illustrated by falls in the average period for an individual shareholder to hold stock from five years in the mid 1970s to under six months in 2006 (Werther and Chandler 2006). This suggests that most shareholders do not seem to want to own part of a company for reasons other than for speculative capital gains. It has led to a fundamental change between managers and owners, meaning that managers now have to focus a disproportionate amount of their time on short-term results, dividend levels and share price which often comes at the cost of long-term strategy, relationships and sustainable growth.

Continuums of corporate governance

Figure 3.1 illustrates the six main continuums of the corporate governance debate discussed above. While CSR ideally demands a shift in corporate governance to the right on each spectrum, it is important to appreciate that there is no one best point on these continuums nor is it desirable to advocate specific points on which firms need to position themselves. The appropriate position for any company will depend on a range of factors which we have discussed earlier. It is interesting to note that a shift to the right is less likely in some countries than in others due to the entrenched differences in the nature of the socio-cultural, as well as political and economic business environments. For example, traditionally, Anglo-Saxon countries such as Australia, UK, USA and Canada have governance systems that are largely positioned on the left of the continuum while European countries such as France, Italy and Germany tend to have governance systems that are leaning towards the right. So in order to do well in a globalised business environment, firms need to be sensitive and responsive to the cultural context in which they do business. Indeed, many firms have been found to increasingly adopt forms of hybrid governance systems that combine elements of both (Makadok and Coff 2009).

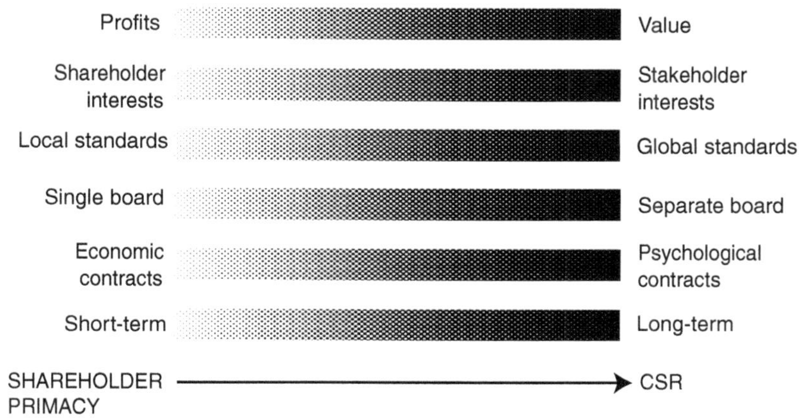

Figure 3.1 Continuums of corporate governance from a CSR perspective

Corporate governance law and its influence on CSR strategy

In developing an effective CSR strategy, there are many 'soft law' corporate governance guidance and 'hard law' obligations to understand and abide with (Robbins and Smith 2005; Sullivan 2005). In addition to the many laws in areas such as occupational health and safety, industrial relations and environmental management, in most countries there are also specific laws, regulations and guidance relating to corporate governance and CSR. For example, in the UK, the UK Corporate Governance Code (formerly the Combined Code of Corporate Governance) sets out standards of good practice in relation to issues such as board composition and development, remuneration, accountability and audit, and relations with shareholders. All companies incorporated in the UK and listed on the Main Market of the London Stock Exchange are required under the Listing Rules to report on how they have applied the Combined Code in their annual report and accounts. In 2001, the UK Company Law Review also recommended that all companies of significant economic size should produce an operating and financial review (OFR) as part of their annual report and in 2004 the UK Government announced new legislation requiring large UK-quoted companies to produce an OFR. An OFR must include information about a company's triple bottom line performance. Similarly, in the USA, the Sarbanes Oxley Act (2002) and revised New York Stock Exchange (NYSE) listing requirements have promoted stricter standards of corporate governance and financial auditing (NYSE 2008), and in Australia companies have to consider compliance with the Australian Securities Exchange (ASX) *Principles of Good Governance and Best Practice Recommendations* (2003), the Corporations Act (2001) and the ASIC Corporate Law Economic Reform Programme (Audit Reform and Corporate Disclosure) Act – CLERP 9 (2004). The Australian system of corporate governance system is presided over by the ASX Corporate Governance Council which, unlike systems in other countries, is not legally binding and set in stone. Nevertheless, more than 2000 companies listed on the ASX evaluate and report their corporate activities against these guidelines. The ASX guidelines lay down ten core principles of good corporate governance (each with its own checklist) which are required to be demonstrated in annual reports. These cover disclosure requirements, relationships between management and directors, ethical and responsible decision-making, and respecting the rights of legitimate stakeholders, etc. The ASX Corporate Governance Principles also explicitly include sustainability in their definition of material business risks to be listed in annual reports.

Despite claims that good governance regulation increases standards of corporate governance, in reality we have yet to fully understand the practical workability and effectiveness of recent corporate governance reforms and what impact they will have in practice. For example, while it is argued that the strength of the Australian, non-binding system is its flexibility to enable different sized companies to adopt appropriate structures and systems, it also allows some firms to potentially employ a tick-box approach. For example, one requirement which is commonly abused is the ASX requirement for firms to have a certain number of independent directors on the board. As Verrender (2009) points out, despite claiming independence, many directors may feel that if they do not abide by the decisions of the chairman that they will lose their directorship role. For this reason, they often recommend employing an independent 'external expert' to provide such advice; a misnomer since no one can be entirely independent if they are being paid for their service. This has given rise to a lucrative market for external consultants who on too many occasions, simply tell the board what they want to hear, thus undermining the impartiality of the role of independent directors.

While Australia's system has been criticised for being too flexible, on the other hand, many of the mandatory systems developed in other countries such as the USA have been criticised for being too bureaucratic, complex and costly to implement and to be a knee-jerk reaction to perceived regulatory failures. In other words, they are driven more by public pressure to be tough on executives than by an informed understanding of the systematic forces that shape board effectiveness. It is ironic that this push to incorporate more 'hard law' could lead to the unintended effect of weakening effective control. For example, Louisot (2009) argues that tighter regulation will heighten the importance of the audit committee, creating the illusion that the only risk is one of non-compliance, resulting in managers becoming blind to the totality of other risks that face the business. Also, McNulty *et al.* (2005) argue that excessive control would lead to the public losing faith in boards as a central mechanism of corporate governance, thus undermining the purpose of governance reform which is to promote confidence in distant investors as to the adequacy of board governance. While some argue that directors cannot be trusted to choose fairly and freely between the interests of shareholders and the community, others argue that legislation may stifle entrepreneurial spirit which is already in danger of over regulation of directors and companies. Indeed, this latter position is taken by the Australian government which has decided that disclosure of triple bottom line outcomes will remain voluntary and that the focus should be on soft 'light touch' initiatives which encourage business participation in CSR and the development of national regulations, codes and guidelines. Having said that, the current financial crisis has changed perceptions about what is required in order to comply with disclosure laws. The resulting increase in regulatory vigilance has made it far more complex for companies to decide what information to disclose and when to do so. So reliance on guidelines and rules may no longer be safe because the disclosure expectations of a 'reasonable' investor today are very different to what they were a year ago. This is reflected in the dramatically increased willingness of shareholders to engage in class actions against directors (AU$400 million in 2007 to AU$1.2 billion in 2008) (Stock 2009).

So it seems that using current corporate governance laws and regulations as a benchmark to develop an effective CSR strategy, is of limited use. Managers must realise that it is the publically perceived standard of moral indifference as well as legal intent that is increasingly used to judge companies these days. To understand this further and therefore manage it more effectively, it is useful to consider Pemberton's (2005) argument that moral indifference stems from the increasing physical and emotional separation between managers and stakeholders in many companies. Pemberton argues that as emotional and physical distance increases, feelings of responsibility and an awareness of consequences decreases. This brings us to multinationals which face special challenges in developing effective CSR strategies, because of the geographically dispersed, remote and international nature of many of their business activities.

International law and multinational CSR strategy

If controversy and difficulty exist in using domestic laws to guide CSR strategy, then the challenges of influencing multinationals in the international arena are much greater. As we have seen, corporate governance is governed by the law in different ways in different countries, leading to very different governance structures and practices. For example, in the USA, power is normally concentrated in the hands of one person who both chairs the board and is the top manager of the company (the CEO). However, in Europe, the separation of the chairman and CEO is standard practice on the basis that the board cannot possibly oversee the effectiveness of the CEO if they are the same person. To complicate things further, some countries have sought exemptions from the strict requirements of the US Sarbanes Oxley Act. So understanding the rights of stakeholders in the operations of multinationals with activities across national boundaries are complicated by these different approaches, as is taking action for any breaches of these responsibilities.

In resolving this problem, one option is to resort to international law but as Tully (2005) points out, corporations are relatively invisible in this context. Currently, international law affirms the primacy of national courts (where firms are incorporated) in settling issues of corporate governance because it is generally assumed that parent companies should exercise control over subsidiary decisions. Furthermore, ethically, it is arguable that subsidiaries should adhere to standards of care at least equal to that of their parent country. However, despite these assumptions, in practice it is still difficult for a parent company to be held accountable for the extraterritorial conduct, and host governments are often reluctant to enforce claims because of the detrimental employment and revenue implications of that company relocating to a more sympathetic regime.

Another problem in holding multinationals responsible for overseas activities arises from two basic principles of company law. The first is the doctrine of separate legal personality between parent and subsidiary firms, and, second is the limited shareholder liability of parent companies for subsidiary activities. Typically these are both upheld by the law despite the fact that parent companies do indeed have the capacity to control the activities of their subsidiaries (Tully 2005). While recently (for example, under the US Alien Tort Claims Act (ATCA) 1994) some courts have been prepared to hold parent companies to account for their subsidiary's actions, particularly when the company as a whole is structured as a unity, it is still rare for national courts to impose liability on domestic firms for their actions overseas (particularly in the UK and Australia). This means that the increasing global reach of companies has not been matched by a coherent global system of corporate accountability.

Soft instruments of governance and CSR strategy

While there is no current international corporate governance law specifically relating to multinationals, there are numerous international 'soft' instruments that have been developed by international bodies which are often used as benchmarks of acceptable behaviour. Multinationals cannot afford to ignore these provisions since they have been drafted by governments who are the principle law makers in the jurisdictions where they are incorporated. So these soft, legally non-binding provisions effectively become a source of hard law when utilised by national courts to fill gaps in their national laws and practices. Examples include the United Nations (UN) Global Compact 2004, the UN Code on Transnational Corporations 1990, the UN Universal declaration of Human Rights 1948, the International Labour Organisation (ILO) convention concerning freedom of association and protection of the right to organise (1950), and the Organisation for Economic Cooperation and Development (OECD) Principles of Corporate Governance 2004, etc.

Another source of governance control are the many Private Voluntary Initiatives (PVIs), such as codes of conduct and practice, drafted by professional bodies, employer federations, business councils and NGOs such as Transparency International, Friends of the Earth and Greenpeace, etc. Although useful as indicators of acceptable moral behaviour, these operate outside conventional government sources of regulation and have no formal legal authority (Mertus 2000; Braithwaite and Drahos 2000). They are also often criticised for being ideologically motivated. Nevertheless, failure to abide by them can still lead to exclusions from associations, deregistration, and damage to reputation. So they should certainly be on the radar of multinationals when developing their CSR agenda.

In addition to the above, managers should be aware of the numerous global sectoral advances that have been made in the construction and engineering sector such as UNEP's Sustainable Buildings and Construction Initiative (SBCI), the Engineering and Construction Industry Anti-Bribery Principles developed under the auspices of the World Economic Forum, and the International Federation of Consulting Engineers (FIDIC) anti-corruption principles, etc.

Finally, there are the many independent certifying and accrediting bodies which produce CSR standards and guidance. These include: the International Standards Organisation which produced Corporate Social Responsibility standard ISO 26000; Social Accountability International (SAI) which developed Social Accountability 8000 (SA8000); and the Institute of Social and Ethical Accountability (ISEA) which produced Accountability 1000 (AA 1000), etc.

The law of tort and CSR strategy

Many of the above are used by companies to benchmark socially acceptable standards of CSR and defend accusations of negligence in countries where the law of tort applies. The law of tort permits injured parties to commence actions against another party who has committed a civil wrong, outside legislation and individual contracts. Very simply, to make a claim in tort a party must prove they have suffered injury, that they were owed a duty of care by the defendant, that this duty has been breached and that the breach has 'caused' the injury. Proof of fault through omission or deliberate intention is not normally necessary – negligence is normally enough.

Under Australian law, if it is not possible to prove negligence on the part of directors then the only way a claimant can bring an action against a company (as a separate legal entity) is if it can be proved that the corporate governance arrangements which controlled those directors' actions were somehow at fault (Tully 2005). This is not easy and in Australia the normal practice of courts (say for environmental offences) is to impose liability upon directors and other corporate officers. Nevertheless, companies can be held vicariously liable for the negligent and unauthorised actions of their employees undertaken within the scope of their employment, unless of course the board can prove that it was diligent and had adequate corporate governance systems to ensure board effectiveness and accountability to prevent that conduct. As we have shown above, this cannot always be assured by merely focusing on compliance. Instead, a company must show that it has based its governance systems on best practice as informed by current research in this area, some of which we review below.

The board of directors

The board represents the heart of a company's governance system and comprises both executive and non-executive directors who are expected to exercise trusteeship on the shareholders' behalf. The function of non-executive directors is to bring an independent source of advice, expertise, experience and judgement to the board. Independent means free from management responsibilities and any other business relationships (apart from fees and shareholdings) which might colour their judgement and advice. The board of directors has a duty to direct the company and to present a clear and accurate picture and assessment of the company's position for potential investors through reports and accounts that present a clear and balanced view (good and bad news). The board is empowered to direct and control a company by a constitution and is responsible to develop and implement corporate strategy. It does this by defining policies for management, setting performance targets and monitoring performance against them, ensuring risks are managed effectively and by ensuring compliance with relevant laws and regulations (Matheson 2004). The board's business should be directed by agreed procedures and it is the company secretary's responsibility to ensure that these are followed and that relevant rules and regulations are complied with. It is important to point out that the board does not manage the organisation but is responsible for employing managers to do that under the direction of a CEO who is responsible to the board for the day-to-day leadership and management of the organisation and implementation of strategy. The important point here is that management and governance are quite distinct activities, although this does vary from country to country. As we have said, in some countries the CEO and chairman can be the same person while in others the roles are separated. Similarly, some countries adopt the 'unitary board' structure, which brings all executive and non-executive directors together, while others adopt a 'two-tier' structure which separate the governance function and the management function into different bodies.

Improving board effectiveness and accountability: the issue of independence

Calls for alternative approaches, particularly in relation to more independent board processes and accountability have surfaced in the years following corporate scandals such as Enron, etc. Questions have been raised about the sometimes 'cosy' relationships between executive and non-executive directors which can lead to collusion and impartiality in the board's decision-making process. Therefore, many recent corporate governance reforms have included an explicit requirement that there are greater levels of independence on boards and that independent non-executives should preside over remuneration and audit committees, where conflict of interests are most likely to arise (for example, Sarbanes Oxley Act 2002 and ASX Guidelines 2003).

This current thinking places emphasis on the ability of independent non-executive directors to balance support for executives in their leadership of the firm, but at the same time, control and monitor their conduct (Roberts *et al.* 2005:s6). However, Shen (2005) cautioned against making the assumption that independent non-executive directors intuitively know how to balance these two roles, and that they can be relied upon to be diligent stewards in performing their governance duties out of their own volition. These concerns are not unfounded especially when there is evidence that, in many cases, executives exert a strong influence on the nomination process of independent non-executive directors. Indeed, even when nomination occurs through external voting, independence is questionable since most applicants are sourced through executive networks which have an interest in having a friendly and sympathetic board to sanction their decisions (Hubbard 2004). Therefore, in many instances, independent non-executive directors are not effective in discharging their governance responsibilities for fear of social sanctions by fellow directors if they challenge executive decisions or participate in actions that threaten the common interests of other executives (Westphal and Kanna 2003). To avoid this, Shen (2005) argues that incentive systems and remuneration processes for independent non-executives also require reform. For example, the potential for any conflict of interests can be mitigated, by ensuring that independent non-executives keep a large portion of their stocks throughout their tenure and for several years after they step down. This helps to ensure that they do not focus on short-term stock prices but on the company's long-term performance. Others advocate limiting the number of independent directors who may serve on many boards at the same time so that they can devote more time to their role in one firm. However, the jury is still out on whether these measures will actually work in the interest of corporate governance reform. It is beyond the scope of this book to discuss this in detail. Suffice it to say, the current corporate governance reforms that demand structural and compositional changes to company boards by simply stipulating the appointment of more 'outside' or 'independent' non-executives have limited capacity to resolve many governance problems (Tosi *et al.* 2003). Indeed, many argue that true improvement in corporate governance will only come when boards have much wider stakeholder representation and we will discuss the practicalities of doing this in the following section.

Stakeholders as directors

We have argued that, today, boards are expected to ensure that their businesses are not divorced from wider society but interact with it and contribute to it. We have also argued that corporate success is increasingly judged on their ability to do this at the same time as maximising profits for shareholders. This requires a shift away from the stakeholder primacy model to considering the interests of a much wider stakeholder constituency. As OECD (2004) recommends, even when stakeholder interests are not legislated, where appropriate, firms should develop stakeholder relations policies and make additional commitments to stakeholders in recognition of broader interests.

There are clearly many different mechanisms which can be used to achieve this objective. The first and most obvious is to carefully consider what type of board companies want in terms of involving stakeholders and how much power to give them. To this end, Hubbard (2004) identifies six different types of board which vary depending on the amount of power given to stakeholders and managers (see Figure 3.2 below).

In *stakeholder controlled* boards, the board of directors comprises a wide variety of key independent stakeholder groups to represent a wide variety of interests and has control of most issues. There are few executives on such

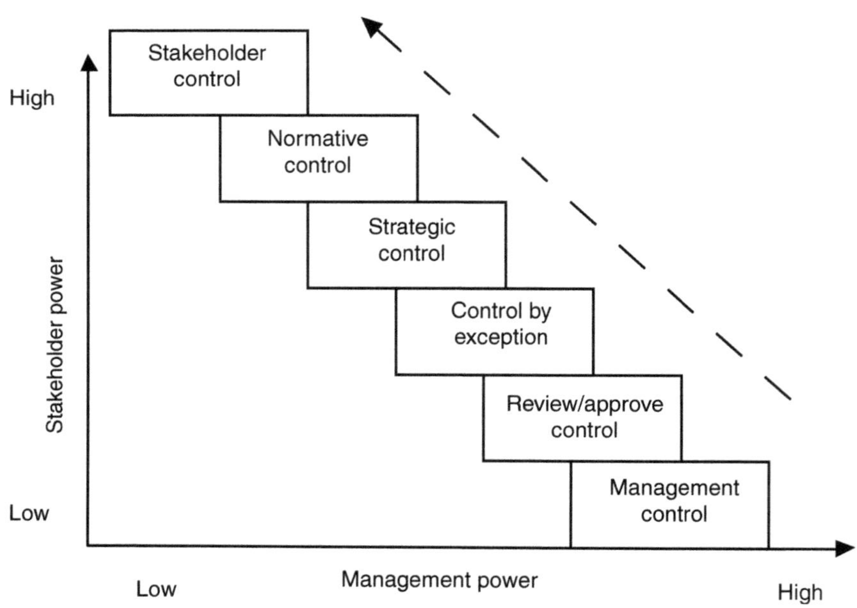

Figure 3.2 Types of board by relative stakeholder–management power

Source: Adapted from Hubbard 2004:289

boards and board approval is required for any major decision. Some NGOs, not-for-profit and government boards have structures like this. In contrast, on *normative control* boards independent stakeholders are well represented but stakeholders retain most of the power and managers can only move within limited decision-making constraints. On *strategic control* boards there is a balance between the power of stakeholders and managers. There is a clear distinction between strategic and operational issues, where the board retains control over the former and managers are delegated control over the latter. In *control by exception* boards, the few non-shareholding stakeholders that are represented have less power than executive management. Stakeholders are able to change management decisions but it is expected that this will be rare. Any proposals put to the board are expected to be approved. On *review and approve* boards, there are few if any independent stakeholders and they hold very little power. The board is dominated by executives and although issues are discussed with stakeholders, the board will invariably approve its own decisions and act in shareholder interests. Finally, *managerial control* boards have no stakeholder representation outside the shareholders who are normally all managers. These boards are rubber stamping mechanisms that perform no real function other than serving the interests of shareholders.

It is important here to emphasise that while some level of not-for-profit stakeholder board representation is good for CSR, boards are full of power asymmetries that shift with time and situation in response to the multitude of contingency factors that are at play in each company. This means that the characteristics of effective boards are impossible to anchor and should always be under review and scrutiny (Pye and Pettigrew 2005). We discuss how this is best achieved below.

Board membership and CSR

Assuming that a model is pursued which provides at least some power to stakeholders, the second step is to decide which stakeholders should be represented and how. Detailed practical approaches for consulting and involving stakeholders in business decisions were discussed in detail in Chapter 2 but here we consider governance issues which can help to facilitate stakeholder engagement at board level. All organisations have different mechanisms for appointing board members, both executives and independent non-executives. Matheson (2004) argues that the best governed organisations do not delegate the task of selecting board members to the chairman or CEO but have an internal governance subcommittee made up of independent directors. This committee should review board membership, establish clear appointment criteria, identify, interview, and undertake due diligence checks and then propose candidates with required competencies for consideration by the board. Nominations should come from any group that is permitted to do so under the company's constitution and some companies will advertise nationally for expressions of interest and use search firms which provide director recruitment services. It is now increasingly common for NGOs to nominate themselves for board membership, and the idea that NGOs such as Greenpeace are better served by working *with* rather than against the interests of big business is gaining acceptance in the NGO sector. Nevertheless, Petrovic-Lazarevic's (2008) research into the corporate governance practices of firms in the construction and engineering found no examples of NGOs, community members or even supply chain business partners on boards of directors. Indeed, none of the firms analysed even had a system for community feedback on their activities. Considering that this research captured the practices of some of the largest firms in the industry (accounting for 42 per cent of annual turnover), it would seem that stakeholder involvement at board level may be good in theory but rarely practiced in reality, at least as far as the construction and engineering industries are concerned.

In bringing about a change in this situation, much can be learnt from community-owned companies which have constitutions that allow some directors to be appointed to the board by the community in the annual general meeting. Similarly, directors might learn from the cooperative business model where shareholders are supply chain partners and a proportion of wealth created goes to the community that provides the resources. A less radical idea is to simply establish temporary advisory boards, management committees, task forces, working groups and workshops to act as a forum for discussion with different stakeholder groups on specific issues or simply to help build relationships with them (Petrovic-Lazarevic 2008). Typically, these types of groups have no delegated authority, liability or legal responsibility and would normally report to the CEO, the board itself or a subcommittee of the board. Although limited power is attributed to stakeholders by these informal mechanisms, when given clear tenures and objectives they can be very useful to an organisation that wishes to move towards CSR tentatively.

Overcoming the risks of liberated boards

We have argued that due to the many social, political and regulatory changes discussed in this book, the role of the board has changed significantly to being about:

- Performance rather than conformance
- Stakeholder interests as well as shareholder interests
- More genuine independence
- Engaged independence rather than confrontational independence
- Long-term performance as well as short-term performance
- Triple bottom line performance rather than just financial performance.

However, as Charan (2005) points out, while liberated boards have many advantages, they also pose considerable risks if not managed effectively.

For example, board outcomes can be compromised by complex group dynamics and interactions can become confused by the balance of power and the number of opinions being considered. Furthermore, boards can become bureaucratic with stakeholders making an endless list of demands from the board for more information and protocols to be followed, distracting attention away from dealing with progressive tasks. Finally, there can be too much focus on insignificant issues putting the CEO in defensive mode, lengthening board meetings unnecessarily and wasting time.

Board culture

Empirical studies of board effectiveness have found that it is not so much the structure, composition and processes employed by boards that are the key to success but a dynamic interaction between behavioural and structural factors (Horrigan 2005). Well structured boards with apparently good processes often come unstuck in their CSR performance because they fail to create an appropriate and conducive culture that supports CSR. This is the 'soft' side of corporate governance which is too often excluded in governance models and regulations. In contrast to the widely accepted notion of organisational culture, board culture is something that is rarely discussed, but is important nonetheless because it represents the values that underpin its actions and therefore those that guide the rest of the organisation. As Charan (2005) points out, every board has unwritten rules that guide behaviour. An effective board culture is one where the rules of engagement are clear and there is a willingness to challenge and question existing ideas and organisational norms in the interests of the organisation and society. The important point is that if this is the type of culture that is present within the board, then it is more likely to be translated throughout the organisation. A strong board culture to which people can relate and subscribe has an immensely powerful regulatory role in achieving CSR strategy.

Cameron and Quinn (2005) propose a useful framework which can be used to understand the relationship between CSR and different board cultures. They propose four main types of organisational culture: *the clan*; *the adhocracy*; *the hierarchy*; and *the market* culture. The clan culture is arguably the most conducive to CSR. It is very much like a family with an open environment of trust and sensitivity to the needs of others. The adhocracy culture could also be conducive to CSR in that it is a dynamic, creative and entrepreneurial place to work which is sensitive and responsive to its business environment. People are encouraged to show individual initiative and freedom and to take risks. In contrast, the hierarchical culture is very formalised, structured, inflexible and driven by procedures. Leaders act as coordinators and controllers and success is measured by efficiency, stability, predictability and dependability. Finally, the market culture is results orientated and the least conducive to CSR. Managers manage by objectives (normally financial) and success is measured against these KPIs, relative competitiveness and by market share.

While useful to understand the relationship between culture and CSR, needless to say, this typology of organisational culture is included as a guide only. No organisation can fit neatly into any of the types, nor is it advisable to do so. It would not make sense to suggest, or indeed expect that firms regardless of age, type, status, or size, can easily switch from one culture to another such as for example, from market culture to clan culture. Importantly, firms need to treat board culture as an evolving concept and understand that it serves different functions at different times. To elaborate this further, when a company is newly formed and is at its initial stage of growth, board culture is more about reinforcing, articulating, and clarifying what the company stands for. Towards the midlife of the company, as it diversifies and grows, boards need more flexible cultures that can make the tough decisions about what parts of the organisation are to grow, contract or even shut down. Finally, as companies reach a stage of maturity or decline, it may be that a complete overhaul of the organisational culture is needed to improve innovation and competitiveness or simply to continue surviving.

Having argued the importance of an effective board culture, using it as a gelling agent to achieve CSR can be fraught with challenges. As Schein (1984) points out, organisations invariably will have multiple overt and covert cultures which are neither easy to identify, decipher, monitor and change. To illustrate, a model of organisational culture (Figure 3.3) developed by Hawkins (1997) shows how the artefacts and behaviours of culture sit at the top two levels while the motivational roots are at the bottom level.

Level 1 – Artefacts
• Policy statements, mission statements • Expressive practices, symbols, expression, rituals and forms.
Level 2 – Behaviour
• What people say and do; what is encouraged and rewarded • How conflict is handled and resolved; how mistakes are treated, etc. • Importance of adhering to existing norms, values and beliefs.
Level 3 – Mindset
• Organisational 'world view' – ways of thinking that constrain behaviour • Organisational values in use, basic assumptions.
Level 4 – Emotional ground
• Mostly unconscious emotional states and needs that create an environment within which events are perceived.
Level 5 – Motivational roots
• Underlying sense of purpose linking individual purpose and motivations to those of the collective • Basic assumptions that determine how individuals think, perceive and feel.

Figure 3.3 Five core levels of organisational culture

Source: Adapted from Hawkins 1997:426

The above demonstrates that an effective board culture is not enough if it does not permeate deeply to affect the motivational roots of peoples' behaviours. It is at this bottom level that culture is most effective in aligning individual purpose and motivations with those of the collective. And hence, it is at this level too, that companies should focus their attention, for when they talk about changing their organisational culture it is more to do with changing the value and belief systems of individuals and not just tampering with the superficial norms, symbols and rituals at board level. And it is also at this level that culture change poses the biggest challenge to firms. Case studies of companies that have made a conscious decision to change their organisational culture have shown that more often than not, a greater proportion of employees at lower levels had only 'adapted' their values slightly, which suggests that any resultant changes in behaviour would be merely short lived (Ogbonna and Harris 1998).

Governance, disclosure and CSR

We finished Chapter 2 on the issue of disclosure of strategy and similarly we finish this chapter on the issue of disclosure of governance. As OECD (2004) points out, a strong disclosure regime that promotes real transparency is a pivotal feature of good corporate governance and is central to shareholders' and stakeholders' abilities to exercise their rights. Although there is an opinion in some quarters that too much disclosure breaches rights to privacy, reduces competitiveness and is potentially bad for business, it is increasingly recognised that transparency and openness will be a critical aspect of effective corporate performance in the future. Perceptions of risk, integrity, reputation, public trust, credit ratings, investor confidence and employee morale are dependent upon employees and external stakeholder groups having access to honest, clear and meaningful information about the companies' activities and how they conduct their business. Ultimately, the decisions about what to report, when to report, how to report and who to report to, rest with the board and have to be made within the constraints of the many laws, regulations, guidelines and most importantly, public expectations that govern what is 'material' in terms of disclosure. Unfortunately, however, Jopson (2007) found that although more companies are increasingly reporting on environmental and, to a lesser extent social impacts, they are only making patchy progress in giving investors and stakeholders meaningful, open and honest information about their performance and prospects in these areas.

Conclusion

Corporate governance is concerned with ensuring sufficient oversight of business to ensure boards of directors obey the rule of law and the interests of customers, shareholders and wider stakeholders in society. Corporate governance has come under the spotlight of government regulators, the public and scholars in the last few years due largely to the failure of existing corporate governance structures and processes to curb corporate misconduct. The construction and engineering industries are right at the centre of this debate and we have proposed in this chapter that there needs to be a more value-based approach to corporate governance which is more sensitive to the needs of a wider range of stakeholders. We believe that a core part of a more value-based governance approach is for boards and top management to take stakeholder interests seriously and focus more on long-term rather than just short–term financial interests. However, we are also careful not to suggest that there is necessarily a best governance approach to achieve this for all firms. The appropriate governance system will differ from company to company and must be tailored to balance the company's own circumstances. Although companies should be guided by codes of best practice and regulations, we argue that they need to go beyond these minimal standards of compliance. As well as its governance structures, we believe that the construction and engineering sector should develop a more ethical business culture as an important basis for CSR, something we discuss further in the next chapter.

4 Strategic business ethics

In this chapter we argue that although firms need to abide by the law, merely doing so does not necessarily satisfy a firm's ethical responsibilities. We also argue that by having certain basic core ethical values to fall back on, that go beyond those required by law, companies have a greater chance of developing and implementing an effective CSR strategy.

Introduction

Business ethics are a body of principles relating to standards of human conduct that have been developed to govern the behaviour of individuals and groups within business organisations. The concept of ethics is inextricably tied to CSR since it is concerned with discretionary behaviour that is not motivated by legal compliance or sanctions, but by moral values and beliefs about what is right and wrong. The contribution that *business* ethics have made to the wider ethics debate is in linking moral and business practices, and in broadening the subject to a wider organisational domain. In firms, it is rare that the flaws of one individual acting alone fully explain corporate misconduct. Rather, as Paine (1994) notes, it is normally the case that unethical business practice involves individuals cooperating within the context of a wider organisational culture which creates the environment for such behaviour. Business ethics therefore acknowledge the role that managers have in shaping ethical behaviour and provide some useful frameworks for making and evaluating business decisions, particularly in ambiguous situations where moral values and organisational imperatives clash (Carter *et al.* 2005).

While business ethics have in the past been a rather peripheral dimension of company policy, these days, with heightened societal expectations, companies are being compelled to publicise or explicitly articulate their ethical stance.

Ethics in business

While there is now a greater need for companies to behave ethically in the eyes of the community, this trend has only grown in the industrialised world fairly recently. Inglehart (1977) put this down to the emergence of *post materialist values in society* where individuals are placing greater stress on social equity, community, belongingness and quality of life. According to Abramson and Inglehart (1992), this value-shift has brought about profound cultural changes that are manifested in the rise of environmental, social and political activism in many societies, and which is now gradually making its mark in the business realm through public demands for more transparent business practices. While this may be the case, business ethics as a discipline is still in its infancy and we have a lot to learn both practically and conceptually (Giacalone *et al.* 2008). Until now, operationalising ethics in practice has been made difficult by the complex philosophical foundations and terminology associated with the area. As Viljoen and Dann (2000) point out, the truth is that most organisations have yet to develop an ethical mindset and culture, despite many having ethical codes of conduct and guidelines to abide by. For example, a recent survey of 14,000 employees found that only 39 per cent thought that their employer had effective processes for managing unethical behaviour and 25 per cent were unsure whether there were effective processes for dealing with workplace grievances (Risk 2009). In the construction sector, ethical standards have been a recurring concern, but independent empirical research into the nature of construction industry ethics is rare. Notable exceptions are Ray *et al.* (1997) who found that cover pricing (deliberately losing a tender), bid shopping (revealing subcontractor tenders to their competitors to force down prices), bid peddling (playing bids off each other), bid loading (loading bids to recoup loses elsewhere) are considered acceptable and common practice. Similar findings were reported by Zarkada-Fraser and Skitmore (2000) who found that opinions about what constitutes ethical behaviour vary considerably across the industry. Ho (2003) found that while many firms do have codes of ethics, many employees fail to understand and implement them because they are not taken seriously and communicated effectively. More recently, Tow and Loosemore (2009) found that three main factors influenced ethical practice in the construction and engineering industries: the absence of ethics education for employees; the absence of rewards for those who act ethically; and the low level of 'process visibility' which makes it easy for people to hide unethical practices. They also found that in reality, most managers in the construction industry appear to still see ethics as a rather daunting, unattainable, soft and nebulous aspiration. These findings are reflected by Fewings (2009) who also found that a widespread absence of ethical codes and policies in the construction industry forces employees to rely on those provided by their relevant professional institutions. Yet Beder (1998) argues that despite the demands of professional codes to place the community interest above that of themselves and their employer, there is very little evidence to indicate that these professional codes of ethics and conduct are anything more than professional 'window dressing'. This is because complaints are rare and professional institutions have very little power to impose serious sanctions on those that flout its rules. According to Beder, ultimately it is the employer or client that often has the greatest power to require and influence ethical standards of behaviour through their control of economic rewards and sanctions.

Promoting ethics in business

The first issue to address in promoting ethical business practice is in understanding the relationship between ethics and business performance. The nature of this relationship is currently inconclusive and there is even debate about whether ethical programmes lead to more ethical business behaviour. For example, sceptics such as Schwartz (1990) offer warnings to those who might embrace the ethics agenda with too much enthusiasm and take it too far. According to Schwartz, in reality the rule of business is the law of the jungle where only the fittest and leanest survive. In many instances, this ensures that companies which abide by strict codes of ethics are often less successful than those which do not. This has been acknowledged as a problem in the construction sector by Loosemore *et al.* (2003) and Dainty *et al.* (2007). For example, unscrupulous companies who illegally employ migrant labour at below minimum wages and who flout laws relating to workers compensation, superannuation, etc. often get awarded contracts before reputable employers that do not, because of the more competitive prices they can offer unwitting clients. Ethical companies are then gradually forced out of business, lowering employment practices in the industry to the lowest common denominator.

In promoting more ethical business practice, it is also important to understand that business ethics means different things to different stakeholders and that there are many levels of ethical expectation against which an organisation and its members may be judged. Similarly, where managers are serving a range of stakeholders, as they do in construction and engineering projects, the dilemmas in attempting to optimise a solution can be very challenging if not impossible. In this situation, ethical decision-making is a process of balancing the different legal, economic and moral issues which inevitably requires trade-offs to be made between the interests of different stakeholders. For example, our case studies in Chapter 5 show that many firms who are attempting to drive CSR in their business are faced with the ethical dilemma of implementing solutions that they consider 'the right thing to do' but in which their clients often see little or no value. CSR strategy raises many ethical dilemmas like this, which is why it is crucial to have some basic ethical framework when developing one. Table 4.1 illustrates some ethical dilemmas that may have to be considered in the development of a CSR strategy for a selection of stakeholders which might be involved in a construction and engineering project.

Table 4.1 Ethical issues relating to different stakeholders in the construction industry

Stakeholder group	*Some ethical issues*
Managers	Remuneration, negotiating, disclosure, political donations/ lobbying, insider trading, takeovers, pensions, audits, board independence, fiduciary responsibilities, compensation, equality, redundancy, exploitation of developing countries, overseas operations standards, corruption and bribery, pollution, working conditions, safety and environmental policies and standards, hiring and firing, conflicts of interest, confidentiality, policy on whistle blowers, codes of conduct, tax shelters, etc.
Clients	Negotiating with contractors and consultants, payment, abatements, corruption and bribery, pollution, working conditions, safety and environmental policies and standards, etc.
Project managers	Negotiating with consultants, subcontractors and suppliers, payment, retentions, dealing with the community, corruption and bribery, pollution, working conditions, safety and environmental policies and standards, hiring and firing, racism and discrimination, etc.
Designers	Impacts of design on communities, ecological footprint, economics of sustainability, ethics of building function (animal testing laboratories, etc.), conservation versus economic development, architecture versus functionality, capital versus life-cycle costs, planning approval processes, etc.
Employees	Honesty, secrecy, loyalty, compliance with policies, intimidation, theft, corruption and bribery, safety behaviour at work, redundancy, fair treatment, harassment and discrimination, moonlighting, etc.
Unions	Negotiating, safety, working conditions, compensation, remuneration, employee rights, communications, environmental polution, etc.
Shareholders	Fair disclosure, insider trading, profit versus non-economic goals, takeovers, etc.
Financiers	Fair disclosure, profit versus non-economic goals, accounting polices, audits, etc.
Property developers	Community interests versus shareholder returns/profit, capital costs versus life-cycle costs, sustainability versus rental returns, pricing (rental/purchase) versus affordability, property advertising (disclosure), selling strategies (bid peddling, auctions), etc.
Customers	Health and safety, environmental pollution, value for money, fair profit, fair disclosure, deceptive advertising, human exploitation in production, keeping promises, gifts, foreclosure, etc.
Distributors	Freedom from undue influence, undisclosed payments, good faith in negotiations, gifts, etc.

Table 4.1 Ethical issues relating to different stakeholders in the construction industry, *continued*

Stakeholder group	*Some ethical issues*
Contractors	Price fixing/cartels, transfer of risk to sub contractors and suppliers versus risk premium, claims, bid peddling, fair price, treatment of workers (pay, entitlements, pensions, working conditions, safety, etc.), migrant labour, corruption, bribery, harassment and discrimination, environmental pollution, dealing with the community, balancing profit versus community interests/sustainability, etc.
Sub contractors and suppliers	Off-shoring production/procurement of products, price fixing, transfer of risk versus risk premium, fair price, treatment of workers (pay, entitlements, pensions, working conditions, safety, etc.), environmental pollution, etc.
Government (planners, etc.)	Compliance with regulations, political donations/lobbying, community objections versus economic development, sustainability versus business interests, etc.

Source: Adapted from Viljoen and Dann 2000; Fewings 2009

Philosophical frameworks for ethical decision-making

While there is no need for managers to acquire an in-depth knowledge of ethical philosophy, a basic understanding is useful in helping decide whether courses of action, decisions and corporate behaviours are ethical or not. Furthermore, ethical philosophies often inform the standards that will likely apply if an issue should get to court, since the law in many countries is based on these frameworks. In a nutshell, the philosophy of ethics can be broadly categorised into two main schools of thought, both of which are relevant to business: *deontology* and *teleology* (Hartman 1998).

Deontology

Associated with the writings of Immanuel Kant (1785) who set out a universal set of moral principles for ethical behaviour, deontological (duty-based) theories are about 'doing the right thing' – a term which Chapter 5 reveals as common parlance used by firms in the construction industry to describe their motivation for having a CSR strategy. Concerned with the rights, duties and welfare of each person in society as an individual, deontology is concerned with goodwill and good intention rather than the consequences of decisions. In other words, an ethical act is one that is intended to do good whether good comes of it or not. So for example, if a manager makes a decision that causes harm to someone out of goodwill it is ethical (even if the consequences are disastrous). However, if a manager makes a decision out of self-interest, then it is not ethical.

Teleology

Teleology is concerned with the welfare of society as a whole and the trade-offs with individual rights that this may involve. In contrast to deontology, the morality of a decision is measured by the outcome of that decision rather than the intention. So if a manager makes a decision which causes harm, it does not matter whether it was made in goodwill or good intention.

Practical frameworks for judging ethicality

Within each of the above philosophies, a number of practical frameworks have been developed by legislators, courts and decision makers to help managers judge whether they are acting ethically or not. This is shown in Table 4.2. Each provides a different perspective and ideally, they should *all* be used as a frame of reference by managers in judging the ethicality of business decisions.

Linking ethical intent to organisational behaviour

In theory at least, the value of the above frameworks is that they ought to provide managers with a basis for intentional and reflective decision-making in developing a CSR strategy, rather than acting on gut-feeling about what instinctively feels right or wrong. However, as Soule (2002) recently pointed out, while useful for judging whether decisions are ethical or not, none of the frameworks above help managers to understand how to make and implement strategies practically. In other words, how espoused ethical values can be translated into enacted values in the workplace. To help resolve this challenge, Soule (2002) argues that four basic conditions must exist: *comprehension; comprehensiveness; specificity; and contextual relevance. Comprehension* means that all managers must be able to understand what ethics means, what is expected of them, and the intended purpose of doing something. *Comprehensiveness* refers to managers adopting a systematic approach to solving morally challenging business problems, and *specificity* means setting some kind of limits around the factors to be considered in making such decisions. Lastly, an ethical strategy needs to be *relevant* to the context in which a business operates and the unique relationship that a business has with its stakeholders and society.

However, research has found that in many construction and engineering firms the link between moral intent and moral behaviour is tenuous at best. For example, Ho's (2003) analysis of codes of ethical practice in the construction sector found that they are poorly implemented in practice leading to the types of widespread corruption and dubious business practices exposed by Transparency International (2005) and discussed in Chapter 1. This raises fundamental questions about the difference between an *integrity-based* and *compliance-based* approach to business ethics, a distinction which has been

Table 4.2 Practical frameworks for ethical decision-making

Practical deontological frameworks	
Utilitarianism	Utilitarianism recognises that in reality all actions have costs and benefits consequences and managers must choose those actions that create the most good over bad. This approach to decision-making seeks, as its end, the greatest good (or utility) for the greatest number of people.
Distributive justice	This framework argues that ethical decisions are those that lead to a fair and equitable distribution of risks, goods and services and encourage cooperation among members of society.
Virtue ethics	Derived from classical notions of character as defined by Aristotle, this framework argues that ethical decisions are those which demonstrate: honesty, fairness, compassion, generosity, modesty, courage, and self-control.
Fairness/justice	In this approach, the basic moral question is how fairly people are treated. Decisions that show inequality, favouritism or discrimination are unethical because they do not obey this basic rule of fairness.
Personal libertarianism	Ethics is measured not by equality of consequences but by whether a person was motivated by equal opportunity for all in making decisions.
Ethical egoism	Ethical egoism is closely related to the concept of self-interest and looking after oneself first. An ethical decision is one which is first right for the individual but which also as a consequence, minimises impact on others.
Rights-based ethics	Rights-based ethics argue that it is unethical to force people to do things against their will. People have basic rights: to the truth, to privacy, to be safe and to what is agreed or promised. An action that respects these basic rights is considered as ethical.
Practical teleological frameworks	
Rule-based ethics	Managers should develop a clear set of rules to lay down principles to guide decision-making and behaviour within their organisation.
The common good	Presents a vision of society as a community whose members are joined in common goals. Individual good is bound to the good of the whole community and any decision that affects the common good is considered unethical.
Categorical imperatives	Managers are acting ethically only when they make decisions outside standards which they expect others in their business community to abide by.

Source: Adapted from Viljoen and Dann 2000; Fewings 2009

recognised in the organisational ethics literature for some time (Paine 1994; Trevino *et al.* 1999; Jacobs 2004; Rasche and Esser 2007; Vandekerckhove 2007). In essence, a compliance-based approach to ethics is purely concerned with reactive compliance with externally imposed laws and regulations (such as GRI, ISO 26000, SA 8000 and AA 1000) whereas an integrity-based approach involves going beyond compliance by proactively integrating ethical standards into organisational culture so that they become intuitively ingrained into the day-to-day organisational decision-making and operations. In reality, of course, organisations need to adopt a combination of compliance and integrity-based approaches, and we discuss these in more detail below.

The compliance approach

The compliance approach to corporate ethics primarily involves the institutionalisation of ethics by the establishment of specific 'task focused' rules and regulations which are controlled and monitored at individual employee level through mechanisms such as auditing and zero-tolerance sanctioning (through disciplinary measures) of individual actions according to the predetermined standards that apply to their role. Trevino *et al.* (1999) argue that most compliance programmes are usually not very good at bringing about long-term changes in employee behaviour since they are heavily dependent on how effective the governing rules and standards are and how effective the implementation, monitoring and control mechanisms are to enforce them. Rasche and Esser (2007) also argue that this approach hampers ethical reflection and undermines employees' abilities to handle ethical dilemmas which fall outside those predicated within established guidelines. Furthermore, being a top-down approach, it deprives people of the right to question established norms and to engage meaningfully in a discussion of ethical guidelines. Rasche and Esser argue that this lack of engagement fails to inspire excellence and moral self-governance and constrains unethical practice to minimum standards and narrowly defined frames of reference. Bourguignon (2007) even contends that from a utilitarian perspective, the management systems which represent the foundation of the compliance-based approach could be regarded as unethical. This is because they aim to prevent dispute and disorder and make irrefutable the managers' view of the world, ignoring the real opinions, interests and needs of employees who work in organisations. This argument is supported by van Wijk (2007) who argues that ethics cannot be achieved in organisations by simply creating a set of explicit rules for people to follow. While rules have a strong instrumental value and can be used to control people, they cannot hope to cover the many different moral dilemmas one may face in business. Furthermore, they fail to recognise the gratuitous basis of ethical behaviour which are peoples' moral values such as truth, trust and honesty, thereby creating asymmetry between the honesty required from employees and the controlling motives of the company in developing the rules. Therefore, as Babeau (2007) argues, creating rigid organisational rules is actually counterproductive since by inadvertently creating the binary concept of the legal and illegal, managers suppress individual transgression and disorder which he argues is essential to organisational effectiveness in today's uncertain business environment.

The integrity-based approach

The integrity-based approach is aligned with the idea of ethical self-commitment, self-governance and trust in the moral autonomy of employees and other stakeholders. In contrast to the compliance approach which assumes that people are motivated by self-interest, an integrity approach assumes that people are social creatures, not merely guided by self-interest but by values, ideals, friendships, peers and self-imposed norms. A person with integrity is someone who would perform an act only after having reflected on how the action fits with their values and as Vandekerckhove (2007) points out, the assumption is that people with integrity will resist opportunism and unethical behaviour.

At the heart of this approach is the establishment of an organisational norms and values system which is negotiated between stakeholders, self-imposed and voluntary (rather than enforced) and is culturally sensitive to the idiosyncratic characteristics, culture and heritage of each organisation. This sounds fine in theory but, in practice, implementing an integrity approach to ethics is challenging. Managers can find it difficult to translate these ideas into concrete, practical management strategies not least because in many companies, the concept of integrity remains abstract and undefined (Rasche and Esser 2007). In contrast to simply developing codes of conduct and rules, as in the compliance-based approach to ethics, the integrity-based approach involves long-term changes to the espoused and enacted values, commitment and behaviours of all employees and stakeholders such as business partners. This is inevitably a slow consultative process which involves commitment, determination and patience on the part of managers and the implementation of initiatives such as ethics training, mentoring and coaching, ethics champions, ethics testing in recruitment strategies and providing organisational processes, communication opportunities and decentralised empowering structures which foster openness, trust and ability to speak up, disagree and report misconduct without fear of retribution (Brown 2005). Unlike the compliance approach where rules and codes of conduct are rigid and easily defined, the integrity approach involves trusting people to make decisions based on their personal values through flexible rules that can be contested, adapted and reset as the organisation evolves in response to its environment (Vandekerckhove 2007).

Pragmatism and ethics

As we have pointed out, in practice the constraints that exist within a construction and engineering firm's internal and external environment sometimes make the attainment of ideal standards of ethical behaviour difficult. These constraints vary from firm to firm, country to country and indeed from industry to industry. For example, we have shown that ethical reaction to bribery varies considerably around the world and that levels of competition and client preferences in the construction sector no doubt influence a firm's ability to achieve the highest levels of moral behaviour. Therefore, like cultural relativism which we discussed in Chapter 1, ethical relativism is also something that has been argued to have a bearing on differing business practices (Robertson and Crittenden 2003). From this perspective, it may be problematic to apply blanket moral judgements without understanding the circumstances involved and the prevailing norms.

While there may be a case for arguing ethical relativism in justifying a firm's behaviour, at a country to country level at least, recent research suggests that this is increasingly problematic since globalisation has facilitated a process of ethical convergence through the creation of what are known as *hypernorms* which represent ethical standards and principles that transcend national boundaries. This phenomenon is especially prevalent when members of different societies need to work together to achieve common goals and many of these norms are increasingly been enshrined in international law through organisations such as the United Nations. In this context, it is shown that hypernorms not only moderate the ethical decisions of individuals, but also moderate the effects of national identity on ethical decisions (Bailey and Spicer 2007). Whether hypernorms exist at industry level between firms in the construction and engineering sector is unclear. The evidence presented in this book would suggest that this is not the case, although some attempts have been made to standardise ethical expectations across the industry. Ethical relativism may therefore be used quite easily by firms to rationalise their behaviour.

Putting arguments of ethical relativism aside, Frederick (2006) suggests that it is important to be realistic and flexible in judging ethical behaviour in developing a CSR strategy. He advocates using the concept of *ethical pragmatism* in devising a successful framework for dealing with the kinds of ethical dilemmas faced by managers on a daily basis. In essence, ethical pragmatism focuses on the importance of emotion, human judgement and intelligence to understand and resolve problems. Ethical pragmatists argue that the problems that firms encounter and the values that guide responses are a function of a complex and constantly shifting interaction between the firm and its political, social, and economic environment. In other words, from a pragmatist viewpoint, morals are constantly changing and contextual, and are products of human experience and social forces which are community rooted and politically conditioned. From this perspective, it is important to make the distinction between an *unethical act* and a *transgression*. According to Babeau (2007), transgressions are necessary for maintaining competitiveness because they can facilitate innovation and allow managers to work effectively. He argues that absolute adherence to the rules can be bad for business and is often an inhibitory obstacle to the realisation of a task. In other words, it is important that organisations create a space for transgression to occur in order to survive in an increasingly uncertain world where it is often necessary to move away from the rules in order to adapt and respond to complex and unexpected situations. The point at which a transgression becomes an unethical act is of course an area of contention which needs careful defining by the managers themselves within the boundaries of the ethical frameworks we discussed earlier.

An ethos-based approach to CSR strategy

The importance of our argument above, is in challenging traditional conceptualisations of business ethics which have alienated many managers by implying an unattainable 'angelic good' that must be attained. We have also challenged the singular focus on increasingly bureaucratic regulations and codes of conduct which are being uniformly and forcibly applied across industries and companies regardless of their own identity, traditions, culture, history, etc. While this may be an understandable knee-jerk reaction to recent corporate scandals and economic crises induced by unethical corporate behaviour, and while there might be some efficiency and cost advantages in doing so, we question this approach. We have argued that business ethics are not about adherence to a uniform set of rules and that such standardised approaches do not help managers personally engage with the idea of ethics and see them as a value-adding construct. We agree with Cummings' (2005) postmodernist view of strategy which argues that business ethics will only become of value to managers when they incorporate elements of both 'individual' (*aretaic)* and externally imposed (*deontic*) standards of behaviour. Like us, Cummings argues that current (modernist) conceptualisations of business ethics, are dominated by deontic values that subjugate people to discreet non-thinking objects which cannot be trusted to act alone and which reduce ethics to something that can be measured and incorporated systematically into a rational decision-making process. In this approach, the influence of 'surface conditions' that can influence people's ethical behaviour is dismissed and judgements about right and wrong are detached from the cultural context, circumstances and traditions in which they exist. This modernist approach is best reflected in the many rhetorical, unimaginative and indistinguishable codes, value and mission statements which pervade the construction and engineering industries. We argue that while such documents may hit all the right buttons for regulators, they cannot possibly be understood, remembered and internalised by most employees or be of value in directing a competitive differentiated strategy.

Like Cummings we advocate a new approach to CSR based on *ethos* as well as ethics. While ethics are inextricably bound up in definitions which relate to wider socially imposed norms, an ethos-based approach is based on a more self-reflective contemplation of a firm's distinct culture and value proposition compared to its competitors. In our view, effective CSR strategy is not just about being perceived to adhere to universal standards of behaviour but is about acting in accordance with one's own personal disposition, abilities, aims and cultural traditions. In practical terms, this ethos-based approach requires managers to not think of CSR strategy as something that is expected of them externally but as something that emerges through a rigorous process of self-reflection, enquiry and discovery which establishes what the firm stands for, its distinctive identity, history, character, genius and the cultural, economic and other constraints under which it works. This may involve confronting fundamental questions about its core business, the resources around which everything revolves, how these resources should be used, where they come from and what needs to be done to make the business socially responsible, ethical and sustainable. Relationship contexts are particularly important in asking such questions and in defining which values an organisation wishes to be aligned with, whose reactions are most important and how it wants to be seen by others. Finally, there is the question of how an organisation recognises itself as being different.

By prompting organisations to ask themselves these types of questions, the ethos-based approach will help ensure companies develop CSR strategies which do not simply reflect latest management trends and so-called 'best practice' approaches, but which provide distinctiveness, reflect real values and which contribute positively to core business objectives.

Conclusion

In this chapter we have shown that while the field of corporate ethics is fraught with complex jargon, theories and philosophies which can make the subject quite impenetrable, there are some simple principles of ethics that can help inform an effective CSR strategy. While we have pointed out that it is important to abide by the law and regulations concerning business activities in general, we have argued that these prescriptive and minimalistic standards should not constitute entirely the practice of business ethics, and nor should they be the only component of an effective CSR strategy. As firms contemplate CSR at the strategic level, it is important to emphasise that merely adopting such approaches and obeying the law does not in any way guarantee that its representatives will behave ethically in their business practices.

5 Case studies

In this chapter we explore, through a number of case studies, how different types of firms in the construction and engineering industries have sought to integrate CSR into their corporate strategy. Since we have argued that there is no one best way to manage CSR strategically, we do not seek to advocate a single approach here. Rather, we try to convey through responses to a structured series of interview questions with senior managers, the different ways in which CSR can evolve and develop within an organisation and the depth and richness of considerations and experiences involved.

Approach and rationale

Using interviews and case studies we hope to capture a balanced and realistic view of CSR strategy. In particular, we explore some of the context-specific inspiration, experiences and challenges each firm faced in developing a CSR strategy, and how contentious issues discussed throughout this book such as measuring return on investment (ROI) and stakeholder involvement etc. were dealt with. This forms an important part of the book because the majority of CSR literature tends to advocate why firms should be involved in CSR without offering much in the way of how this is done in practice. We also want to convey the gap that might exist between theory and practice which should then allow readers to form their own opinions about where the CSR debate is realistically heading.

We selected the firms for our case studies, not on the basis of their CSR records, but to reflect the range of companies one finds in the construction supply chain. We believe we achieved this but also recognise that due to space constraints there are many not represented here and that there is a lot more research needed in this area. The case studies include: multinationals such as the Lend Lease Corporation, The GPT Group, Leighton Contractors Pty Ltd and Stockland; international designers such as Arup and HASSELL; project management consultants such as Crown Projects; medium-sized family-owned builders such as Kell and Rigby Ltd; and suppliers such as CONCENTRIC Asia Pacific Pty Ltd. We collected our data using semi-structured interviews with managing directors/CEOs and senior

CSR executives from each company. Where relevant we also used information from annual reports and other corporate documents to supplement the interview data. Our intention here is not to promote any company or capture the diverse range and complexity of CSR initiatives pursued. As this chapter illustrates, some CSR strategies were quite limited and simple while others were highly complex and multidimensional. We refer readers to the websites and annual reports for the many details we have had to omit. Nor are we interested in reporting the successes or otherwise of any particular CSR initiative or strategy. Instead, the majority of what we report here is data from the interviews, where we asked the following key questions revolving specifically around the connection of CSR to strategy which is the focus of this book:

1. What are the drivers of CSR in your business?
2. Describe your CSR strategy?
3. Who was involved in the development of the strategy?
4. Is the board able to effectively support CSR?
5. What are the broader external barriers to adopting CSR?
6. What are the internal barriers to adopting CSR within your company?
7. How have you developed a CSR culture?
8. What are the main challenges in successfully implementing CSR strategy?
9. How is CSR ROI measured?
10. How are CSR successes reported?
11. How are CSR failures reported?
12. From a strategic viewpoint, what are the key determinants of CSR success?

The intention of our interviews was not to constrain discussion but to facilitate an informal conversation about each company's experiences of CSR. To capture the richness of the journey experienced by each company we allowed respondents to move outside our question frames and the result was a rich, qualitative and relatively unstructured account which inevitably differed from firm to firm. In our analysis of this data, we decided to present our results in a narrative story rather than attempt to convert the data into some quantifiable code. We are at pains to point out that the narratives are presented purely from the perspectives of the firms interviewed. We have simply reported what was said although to ensure accuracy, we have cross referenced this to annual reports and other relevant documents and sources of information. Ultimately, the veracity of any claims made in interviews or annual reports is for company stakeholders, governments, readers and researchers to judge.

HASSELL

Founded in Australia in 1938, HASSELL is a single, privately owned international network of design studios which employs over 3,000 people worldwide. A multi-disciplinary design practice, it is structured around the key disciplines of architecture, interior design, landscape architecture and planning, with integrated sustainability and urban-design capabilities.

The drivers of CSR

The main drivers of CSR in HASSELL are its company values – 'CSR is not a driver that measures the success of our business – it is the driver that reinforces the values of our business which are about betterment of the built environment'. CSR is also internally rather than externally driven – 'CSR wasn't a response to external demands that we now have to do this because everyone else is doing it. The emergence of the term CSR hasn't and won't change what we do. We do it anyway. But it does provide a framework for us to understand, celebrate and report what we do more effectively. It's not a marketing-driven thing'. An internal report put to the board in August 2008 entitled 'Corporate Social Responsibility (CSR) – a path for HASSELL' does not define a detailed CSR strategy but it does identify HASSELL's values, culture and vision. Its stated values are 'excellence, collaboration, integrity and reward', its culture is 'founded in the stewardship of the land and the creation of exceptional sustainable places for communities' and its vision is 'creating exceptional places'. Its mission is articulated as, 'We will deliver innovative planning and design solutions through our commitment to: a collaborative teamwork approach; leadership and initiative; partnership with our clients; and environmental and social responsibility'.

While internal factors are identified as the primary drivers of CSR, there are also acknowledged external pressures to consider it in a more strategic way. For example, there are increasing numbers of clients who seem to want HASSELL's values to align with theirs and who ask if there is a CSR strategy – 'It helps relationships with certain clients but we haven't done it because of clients – it's not a defining factor. We have done it because it's part of our culture. The drivers are more internal than external. ... We do seek to be a commercially successful business, and we are, but our drivers are much broader than that'. It was also acknowledged that HASSELL is seeking to define CSR in its business more clearly because the wider corporate world is doing it: 'We were doing a lot but we weren't telling anybody so we are trying to measure and define more carefully and accurately what we do through the activities of the Sustainable Futures Unit. It's become something by which corporations are measured'. While HASSEL's CSR strategy is values-driven, an internal report put to the board in August 2008 also recognised the potential competitive advantage that could be gained. It pointed out that many of HASSELL's clients have strong CSR programmes, that few of its competitors promote CSR and that its many existing CSR activities receive limited promotion. Specific reasons put in this document for developing a more structured approach to CSR included: an ethical basis to contribute to the creation of sustainable places; recognition that considering broader stakeholder interests can have significant benefit for HASSELL's long-term financial position and sustainability; enhanced reputation; improved recruitment and retention of high-calibre staff; recognise and build on the existing strong culture of social responsibility and projects underway; the opportunity to demonstrate leadership by being the first Australian (and one of the first international) architecture, planning and design firms to announce a CSR policy and programme.

A final driver of CSR in HASSELL are the firm's employees – 'We were being questioned by some clients but we were also being questioned more by our own people. A desire exists in all parts of our organisation to do the right thing and to contribute to communities in a way consistent with the company values and at the grass roots level to focus on the people in the company'.

Describe your CSR strategy

HASSELL's CSR strategy is still evolving – 'We did some research about what CSR is and through brainstorming sessions we identified two things we wanted to do: restoring our environment and building our community. Out of these emerged two themes: relieving stress and promoting healing (social and environmental); and expanding our horizons (capitalising on the creative side of what we do). This is the basis of our strategy which is still evolving'. An internal report put to the board in August 2008 identified some possible activities and programmes which could support these two themes although this has not yet been formalised into a documented strategy. Possible activities, under the first theme of 'relieving stress and promoting healing' included contributing to: access to food, water, shelter, education, employment and services; community development projects; the creation of healthy communities; the establishment of environmentally sustainable practices within communities; the revitalising and renewal of communities and their environments; and encouraging best practice in planning and design of sustainable communities. Under the second theme of 'expanding our horizons' possible activities included: investing in education, science and the arts to further build our fields of endeavour and our contribution to and creation of sustainable communities; taking our creativity and our knowledge leadership to the wider community and partnering with key groups, organisations and institutions to meet the needs of existing and future populations.

Who was involved in the development of the strategy?

HASSELL has thirty-one owners who elect the board and managing director on a three year cycle. The board appoints three subcommittees which carry out its activities and develop the strategic plan, namely: the executive committee; the operations executive; and the practice executive committee. The practice committee comprises the head of each discipline plus support functions such as HR and drives the more qualitative side of the business which includes CSR. Over many years, the owners of the business appear to have advanced a CSR agenda informally and 'unknowingly' in many ways – 'We regularly meet as the business leaders and owners and talk about these things but it's been loose rather than structured. This current process of putting our CSR into a coherent framework started in August 2008. It's not really a strategy but just a process of bringing it together and reporting it more effectively'.

The main governance mechanism to drive CSR in HASSELL is the recently established 'Sustainable Futures Unit' which reports to the board through the Practice Committee – 'The Sustainable Futures Unit has the task of capturing what has been done for many years, helping us understand what we are doing about this, what people commonly refer to as sustainability and what it means to our business and how to move it forward from a strategic and policy point of view'. HASSELL has also established sustainability working groups in each studio which report to the Sustainable Futures Unit and their recommendations were put to the board in an August 2008 report. This report, which was broadly endorsed by the board, recommended the creation of a CSR policy statement, a sustainability homepage with CSR as a subset and promotional literature, a corporate responsibility manager or shared position between HR and communications and a defined CSR budget – 'We are quite a large organisation which is regionalised and the regions run with a high degree of autonomy and engage in activities individually and there was quite a lot happening already. But as we grow we need to ensure that we do things consistently across the organisation'.

Board support for CSR

The board has strategic focus and issues are brought to the board through the three subcommittees – 'how effectively the board deals with this depends on how it's reported and the response of the board therefore varies. CSR is a values-based thing and inevitably there are some board members that are passionate about it and others that are not as interested – this applies to any company. From the bottom up CSR is working really well. We have a high level of interest within the organisation in CSR because it is our raison d'être (projects that make a difference) but the Sustainable Futures Unit report in 2008 was about trying to get the board to think more strategically about it on an international basis. The board runs issues-based workshops every year and a coming review of the Sustainable Futures Unit will fit into one of those'.

Barriers to adopting CSR

The barriers are largely internal – 'We are a project-focused organisation and the project is king. While we are ambitious about doing a lot of other things, it is the pressure on the project that drives us. So ensuring that we are keeping a focus on things beyond the project is always a challenge'.

Regulatory barriers are not perceived as a problem – 'We are not a public company. We are more interested in whether we are doing enough in our own terms than whether we are doing so comparatively. For us it's not a regulatory or reporting issue'.

Developing a CSR culture

CSR sits as a subset of HASSELL's wider culture – 'it's not really defined to be honest. It is believed but not defined. CSR was a random culture but now it is becoming a broader culture through the work the Sustainable Futures Unit is doing to capture and formalise it more. We do this by values-based workshops, by working on strategies and feeding them through conferences – it occurs by osmosis as much as anything else. We also employ people that fit with our culture but not specifically with sustainability or CSR criteria. But we do make judgments'.

Challenges in implementing CSR strategy

Numerous barriers are identified including board support, understanding what it means for the business, a project-based culture, growth and size and achieving consistency – 'You have to bring the owners along with the strategy. ... When you have a project-based culture, big picture strategies can often not get the attention they should. The only way to manage this is to draw on people in the business who care about these things and draw on their enthusiasm ... We have also moved into a different company model. We are an international company and quite large now so the previous models of doing things don't work so well. ... Fostering some consistency is a major issue when communicating with external organisations ... As a consultancy and industry we are very cyclical – so when things get really tight its harder to deal with these things ... Knowing what it is and measuring it is also quite difficult. There are a lot of people in our business that give their own time and this is not measured or transported into the CSR realm. Understanding what it is and then defining it for your culture is really difficult'.

Measuring ROI

There are deliberately no formal processes in HASSELL to measure ROI from its CSR activities – 'It's not measured. Knowing what it is and measuring it is quite difficult. The systems that are required to do it are so onerous that it will take the fun out of doing it, particularly at the local level where something might come through the door and people will say – hey look we can really help these people. We are currently involved in an initiative called Common Ground where we do work for no fees for the homeless and there are a whole range of initiatives that our staff engage in. That is really difficult to measure. If we measured ROI on this it would really kill it. The key is to measure how effective we are in these many areas without killing initiative. So a lot of it is about judgement'.

The criteria used to assess whether HASSELL supports a cause are largely subjective – 'We may say let's just do it because it's a good thing to do. For example, we are doing some work for the Salvos at cost at the moment but we don't have any strategic partnerships with organisations like this. In the future we may be more focused than we are at present. But at the moment it's often responsive. It happens on the margins of the business although we're trying to put more structure around it to understand and communicate it a bit better but without changing the other things that are happening – we won't ask people to fill out forms because we don't need to – we are a collegiate partnership'.

The internal report put to the board in August 2008 which addressed reporting of CSR argued that better recording of what the firm does and the expenditure involved are important for the following reasons: sound corporate governance and financial management; preparation of Sustainability Reports; celebrating and promoting achievements locally, nationally and internationally; and for recording the internal and external benefits it brings'.

Reporting CSR

Currently there is no formal reporting of CSR activities – 'We haven't had experience of this yet. Starting this year, CSR is to be reported by the Sustainable Futures Unit to the board through the Practice Executive by a regular sustainability report on selected GRI criteria and something tacked onto the financial software and reporting which we have just restructured'. Like many companies, HASSELL has been very good in its financial reporting and is learning about reporting in other areas – 'We are now going through the process of developing new ways of reporting around our CSR objectives'.

The main focus of any future reporting in HASSELL will be internal rather than external – 'We will pick a few GRI headings for the annual report but more importantly we will be reporting to our own people. How they are getting involved, what their achievements are, etc. There is no reason why we won't report failures as well, since we want to learn'.

Key determinants of CSR success

The main determinant of success identified by HASSELL is ensuring CSR satisfies its values and that it helps achieve internal satisfaction and recruit good people – 'It goes back to the issue of our culture and how CSR helps to define us as an organisation and how it helps us gauge the kind of people we want to work with (who share common goals and interests). We want to be doing something of value to those organisations we are assisting through our CSR. It must add value not just for us but for our partners. We would prefer to give advice rather than dollars because it adds more value'.

Arup

Established in 1946 as a consulting engineering business in London, Arup has grown into an international consulting firm which is owned in trust for its employees. Employing over 10,000 people worldwide, Arup's wide-ranging portfolio of projects is served by over ninety offices across Europe, North America, Africa, Australasia and South East Asia.

Drivers of CSR

The main drivers of CSR in Arup are: reputation, minimising externalities, promoting company values, transparency and competitive advantage. 'We have a long-standing reputation for delivering sustainable outcomes for the built environment. Our core values, around which the organisation originally formed, also very much talk to sustainability, and CSR is a part of this. We do a lot of good work and needed to report what we did better but also to up the ante in our performance. We also wanted to reduce our footprint. Being a service-based organisation this is relatively small but we have a major impact in the advice we give to clients. It's also about competitive advantage, about having sustainability integrated into all of our business lines so we avoid green wash'.

In describing the drivers of sustainability, Arup repeatedly emphasises the advantages of independent ownership – 'it gives conviction a place in its decision-making, alongside the needs of clients and commercial imperatives. The result is clear-sighted, thoughtful decisions about its priorities as a business and as a member of society'. Arup's values are still strongly connected to the founding principles of the firm's founder Ove Arup – 'Honesty, fair dealings with others, and a commitment to humanitarian aims were articulated as key principles in 1970 by the firm's founder, Ove Arup. These values remain fundamental to the way Arup operates today'.

Arup's annual sustainability report states that financial viability is one of the key drivers of its strategy – 'We have a balanced approach to, and a business case for all of the work we take on when considering sustainability at Arup, and the sustainability of Arup. We must meet our needs today without compromising the future of the firm. We can see that the key to the future is focus, balance and scale. If we focus on our areas of strategic intent, take a balanced approach to adopting solutions and maintain appropriate scale in our business operations, we will continue to deliver sustainable outcomes for many years to come'.

CSR strategy

Arup do not use the term CSR, instead preferring to incorporate this concept into a corporate sustainability strategy. The strategy is to embed sustainability into all parts of the business rather than separate it out into a distinct function.

Arup's strategy is led globally but operationalised regionally and has four components: its business; its people; its own facilities; and its external relations. 'At the core is our business. In terms of competitive advantage this is about embedding it into all parts of our business that we offer to clients. The next part is our people. Obviously, our business is our people and our asset, and this is about ensuring that the people in our business have an understanding of sustainability and can talk to our clients about it. The next one is our facilities which are our buildings. To make sure we can walk the talk. This is about monitoring the performance of our buildings and their impact and carbon footprint. The last part is our external relations. How we promote what we do. But more importantly about the pro bono and in-kind work that we do which are a very active part of the business'.

The Arup website identifies a whole range of community causes which the firm supports. Much of this is coordinated through 'The Arup Cause' which provides structured opportunities for its people to develop personally, deploy their professional skills and to contribute to development globally in response to the immediate needs facing humanity across the globe. There is also 'The Ove Arup Foundation' which is a registered charity established in memory of Sir Ove Arup. Its mission is 'the advancement of education directed towards the promotion, furtherance and dissemination of knowledge of matters associated with the built environment'.

Who was involved in the development of the strategy?

A core team drives Arup's sustainability strategy at a global level. This is chaired by a board representative for sustainability and has a representative from each region. There is one sustainability policy set at board level which drives the strategy in each region and each region has a certain amount of discretion about how they want to operationalise it. The annual report states that, 'We launched our Sustainability Policy in September 2007. Owned by our Group board and born out of staff engagement, it calls for a systematic approach to sustainability across all areas, highlighting our objectives and outlining the actions that we will take in four key areas: our business, our people, our facilities and our external relationships'.

Arup is not a hierarchical organisation so there is not a complex hierarchical reporting structure. There are regional boards and in each region there is a board representative that is responsible for driving sustainability. In each region there is also a sustainability coordinator who is responsible for delivering the strategy in the region. The annual report states that, 'The Policy provides a shared language to talk about sustainability across the firm. Following the launch of the policy, each of our regions prepared a strategy that addresses the four elements of the policy and their implementation in their area. This is the key to making sustainability relevant across the firm. Each region has clients, cultural issues, laws, priorities and ways of working that are specific to it – the sustainability approach must accommodate this diversity'.

Board support for CSR

'People obviously have very different opinions about certain aspects of sustainability. As always there are different views on the board which is natural in all organisations. There is overwhelming support for what we are doing. The degree to which individual members get involved varies but we don't depend on the group board to get things going in the regions. The impetus for this comes from the regional boards. Different regions are in different places on the journey and we have all tackled things differently. We are not a command and control organisation'.

External barriers to adopting CSR

The main external barrier is clients – 'our main barriers come up at project level where our client may be unwilling to go so far, not willing to go at all or do not quite get it. A lot of work we have been doing is to empower people in the organisation to be convincing about the benefits of doing it. More often than not there is a business case and it's up to us to convince the client it is worth doing'.

Regulations are seen as a minor barrier – 'There are certain aspects of regulation that don't help us to get along but that has never stopped us. It really depends on an individual's level of courage and what business levers we can use to persuade a client what the business case is. You can't ever let legislation get in your way and in terms of competitive advantage clients expect us to be ahead of compliance by quite a long way'.

The barriers also depend on the stage of the project – 'We are very conscious that different parts of our business are able to influence sustainability outcomes to different levels. For example, if you are down at the construction end then your ability to influence sustainability outcomes is constrained – it may be down to materials, etc. Whereas if you are further up in design and planning you have better influence on what gets built and how it gets built'.

Internal barriers to adopting CSR

Internal barriers include senior management support and overcoming resistance to change – 'You need to have board support first and foremost. You need to engage your board so that they know what you are talking about. It's like any change programme. You are going to get resistance to what you are doing. There are always people who are not open and not willing to learn. That is always going to happen'. Arup seeks to overcome this by setting milestones – 'so that it becomes clear what is happening and what is not happening. We can then identify the problem points and try to understand what and why it is not working and what else we need to do to enhance understanding of why it may help their business. You need to keep finding out where the pockets of resistance are and unlocking them'.

Developing a CSR culture

Arup's annual sustainability report states that, 'Formalising something that is so inherently a part of who we are is challenging. To be successful, our approach must work for everyone. It must be relevant to our staff and to the wider community. It must be true to our values and, most importantly, it must continue to make sense for our business'.

The company's values are repeatedly referred to, in describing its culture – 'It goes back to our core values which are very much about social responsibility, which attracts the type of person that drives the culture. So there was already a culture going on. It's not a step change for Arup. We are not like a contracting company that has always been finance driven – that's a huge change'.

The culture is also related strongly to Arup's corporate structure – 'Arup is a trust, we are not owned by anyone else. The employees own the company and all the money that we make gets funnelled back to employees'. Arup's website also reinforces this point. It states that, 'Arup's commitment to a sustainable approach to all its projects is both enshrined in a formal sustainability policy and embraced personally by the individuals that together make up the firm. Arup's ownership structure actively reinforces this approach and holds the firm accountable to its own people for its independent approach and to its social and corporate responsibility'.

We are not beholden to anyone else and there is a huge amount of independence. The annual sustainability report states that, 'Our independence allows us to set our own direction, creating a special environment for everyone to develop to their full potential and to feel empowered to influence, shape and action their ideas, aiding us on our journey to become a more sustainable Arup community'.

Promoting the culture is mainly done through integrating the firm's values into every day decisions – 'We don't have a performance management system that links rewards to CSR objectives. It's just expected. It's part of our structure. Arup talks about the organisation having 'reasonable' prosperity – we are not about making 20 per cent profit at any cost. It's a different mentality to many firms'.

Challenges in successfully implementing CSR strategy

Maintaining momentum and continual improvement are the biggest challenges for Arup – 'One of the biggest ones is maintaining the rage. It's keeping up the pressure and keeping it going. It's about getting everyone talking about it and offering it to clients as part of what we do – like it or not. And it's also about understanding when we have got there – especially when it's not our core business. And then what's the next thing that we do to keep pushing it forward to keep it at the edge and keep it relevant'.

Measuring ROI

Arup sets sustainability KPIs (key performance indicators) but is in the early phases of doing so. In the annual sustainability report the sustainability objectives are arranged and reported against specific KPIs under the four components of the sustainability strategy – '... they will get more sophisticated as time goes on and as we bed things in – we have already had our first review of what they are. We look at things like repeat clients, are we talking about sustainability objectives with clients, etc. Eventually we will be able to link sustainability objectives on a project with things like repeat business ... We do measure all our pro-bono time and we have budgets for our in-kind support. But ROI is hard to capture. We have a very deliberate decision process of who and what we will support. The criteria we use are linked strongly to Arup's business objectives and our peoples' general interest areas'.

Reporting CSR

Early in 2008 Arup released a Sustainability Statement which reported a set of performance metrics for the period 1 April 2006 to 31 March 2007 against the new framework set out in its Sustainability Policy. It states 'Reporting against these metrics influenced our thinking about how to measure our progress during the following year. Preparation of this report has made us realise that there is more we need to be accountable for. As we developed our regional strategies this year, we also developed and included additional performance metrics to more accurately measure our progress'.

Arup's new annual Corporate Sustainability Report brings together two previously separate reports – the Sustainability Statement and Financial Statements. This detailed report outlines the values of the business and exemplar projects, and reports achievement against KPIs in each of the four aspects of Arup's sustainability strategy. There is also internal reporting against KPIs – 'It's not a stick but it's meant to identify areas where we need more work and try to bring it along'. There are four areas of KPIs: 1) Our business (KPIs include: provide value to clients by building upon our reputation for integrated design and a holistic approach to projects; deliver projects recognised for their sustainability credentials, in line with client expectations; evaluate projects with respect to their sustainability risks and opportunities, and, where appropriate, discuss these with the client; and achieve performance that ensures the firm's economic, environmental and financial viability); 2) Our people (KPIs include: employ and retain staff who have a high degree of awareness and expertise in sustainability for all disciplines practised; provide continual education and training for all staff on sustainability issues relevant to the firm's businesses; and support innovative approaches to implementation of sustainability strategies on projects); 3) Our facilities (KPIs include: maintain management systems to assist with implementation of sustainability objectives; aim to use resources efficiently and to minimise waste, usage of water, energy and other consumables in the office environment; develop a strategy to move towards minimising carbon emissions in our operations; endeavour to prevent pollution within the scope of our activities; and develop a strategy for the firm to move towards sustainable procurement); and 4) Our external relationships (KPIs include: partner with organisations that practise sustainability and that enable the exchange of ideas and the promotion of sustainability leadership across their businesses; and fund and work on community projects that achieve the sustainability goals of the goods and services used in our operations).

Key determinants of CSR success

'You have got to believe in it. If you don't genuinely do so, then people will cotton on quickly ... You also need your board to believe in it and support it and provide the necessary resources to achieve it ... You also have to understand the community you work in. You have to know what it means for your business and how it will benefit in terms of competitive advantage'.

Lend Lease

Listed on the Australian Stock Exchange, Lend Lease is one of the largest international integrated property companies. Operating in more than 30 countries worldwide and employing over 10,000 people, Lend Lease offers multiple services across the entire design and construction project delivery process.

The drivers of CSR

Lend Lease's CSR strategy falls under the umbrella of its sustainability strategy and is driven by its values which are defined as: 'Respect, integrity, innovation, collaboration and excellence'. These values can be traced back to Lend Lease's founder Dick Dusseldorp – 'There is a rich history. We didn't start with a blank page. Sustainability has been an integral part of our culture for more than 50 years. Today, our employees insist that making a difference in our communities, improving health and safety standards, and reducing our environmental impacts are central to our business strategy. ...

A socially and environmentally responsible approach to business underpins our operating principles – a lasting legacy of our founder. We believe it's the right thing to do. But it's also the smart thing to do, because it creates long-term commercial value and reduces our operational and financial risk'. More recently, the sustainability strategy has been driven by a new group CEO 'who is very adamant that a balanced score card informs the factual information that leads to the company's strategy'.

In terms of the balance between the economic, social and environmental aspirations in Lend Lease's sustainability strategy, economic drivers still dominate – 'because of the understanding and everyone is used to thinking in this way'. Environment is second – 'because of trends internationally and the pressure to respond'. Social drivers are significantly behind – 'we continually respond to information regarding employers which are obviously important but information regarding stakeholders in the community is by far the least developed ... At every board meeting there is an environmental and social update given and there is more engagement around social issues because people can relate to it. But it's not given the same strategic weighting and emphasis because there are no data, reports and information. We need a language, we need metrics to measure and report performance but on social this is still developing'.

Other identified drivers of the sustainability strategy are Lend Lease's 'employees and the people and companies who are attracted to work there. The catch cry that our employees would always say is that 'it's the right thing to do'. You can have all the strategies in the world and all the drivers but the crucial step is engaging your people'. The website states, 'Lend Lease founder Dick Dusseldorp built a sustainable company on the talent of Lend Lease people. He strived to inspire and engage talented people by creating a work environment where they could flourish ... Global employee data collection will be extended to provide more annual information on workplace diversity, employee learning and development, succession planning and other recruitment issues affecting our employees around the world. This data will be used to shape best-practice strategies for the good health of the organisation'.

Lend Lease is also driven by a realisation of the economic benefits of CSR. On the website there is a quotation from Rod Leaver, Global CEO of Lend Lease Investment Management, who states that, 'At Lend Lease we have long been aware of the competitive advantages associated with sustainability and responsible property investing; however, it is also increasingly becoming a key subject of interest to our investor'.

Describe your CSR strategy

Lend Lease is four years into a five year strategic plan – 'We did not start with a strategy in the textbook sense. We are quite the opposite – we are grounded traditionally in our employees and historical values and only recently looking at the strategic value of engaging in CSR'. The website states that 'Our goal is to be a global sustainable leader'. The sustainability strategy is driven by twenty-three aspirations (economic, social and environmental) developed by a Global Sustainability Group – 'We developed our sustainability aspirations to define long-term objectives against which we can track our progress and report our performance in the consistent delivery of environmental and socially sustainable outcomes in the places we create – as the builder, manager, owner, tenant or investor'. Each business unit is expected to interpret these global sustainability aspirations into plans that fit the diversity of their market conditions and business operations.

If the strategy had to be encapsulated in one sentence – 'it's essentially about imagining a different world ... Our strategy is to set pointers to this imagined world, to set markers on the hill as measurable targets, to help our business units get there and to measure how our businesses are developing ways of getting there'. For example, there is an aspiration that all projects will be zero energy, water and waste – 'We then work back from that and help to understand what that means for the project or business unit and then how do we get solutions to deliver that'.

The strategy is very much about helping each business unit change their perceptions and challenge their expectations – 'Three years ago when we put that out there, people couldn't imagine the world like this, and there was incredible resentment but now every project real-time is achieving these aspirations and people are understanding what it takes to get there'.

Being a global company there is of course significant variation in regions, in the extent to which this is achieved. Every region has to have different lead and lag indicators – 'For example, the UK market has a very different set of drivers where everyone has to be compliant with numerous regulations such as zero carbon whereas in the US it's much slower because only this year are the US passing bills around carbon and energy efficiency, etc'.

Given that Lend Lease is four years into a five year sustainability plan, the aim for next year is 'passing the baton, collapsing the global sustainability group so functions like human resources, communications, health and safety, and risk, understand their role in sustainability. We want to see sustainability embedded across the organisation'.

Who was involved in developing the strategy?

'It was driven originally by the CEO appointing a global sustainability manager whose job it was to pull a global sustainability plan together'. In 2007 Lend Lease established a board sustainability committee which had the task of establishing the strategic direction. The board sustainability committee is advised by a global sustainability group which initially analysed statements and annual reports that had been made around the world by Lend Lease CEOs over the years. From this rich history it created objectives and aspirations and consolidated those into twenty-three aspirations – 'which reflected things that had already been said in the company in its history'. The board sustainability committee formally signed-off those in 2007 and Lend Lease has followed up by establishing KPIs and business plans within each of the 16 business units to reflect these aspirations. All business unit sustainability plans are approved by the Global Head of Sustainability. Lend Lease then consolidates this into overall performance and reports it in a way which is heavily influenced by the requirements of the Dow Jones Sustainability Index and the Global Reporting Index.

The sustainability hierarchy in Lend Lease starts with the board. Under the board are various groups and committees including the Board Sustainability Committee, executive management team (headed by the CEO and including the Global Head of Sustainability) and a Global Sustainability Group (headed by the Global Head of Sustainability) which is a policy setting unit responsible for implementing the strategy and putting systems and processes in place for measuring and reporting sustainability. The website defines the role of the Global Sustainability Group as 'to ensure good corporate governance and set clear guidance on the relationship between corporate policy, global business policy and regional business unit guidelines or statements of commitment'. The Global Sustainability Group is also responsible for establishing the sustainability policy framework for the development, review and approval of any new or existing sustainability-related policies within the organisation. Under the framework, the Global Sustainability Group reviews the development and implementation of sustainability-related policies for all Business Units. Internal and external advocacy is also an important function of the Global Sustainability Group.

External stakeholders are not directly involved in the creation of the sustainability strategy but are indirectly involved through Lend Lease community networks at project level and by Lend Lease's involvement in groups such as the Green Building Council, World Business Council for Sustainable Development, World Economic Forum, etc. – 'Our businesses in every community run community-consultation processes continually. That is part of their remit in getting buildings built. How does it inform strategy? For example, in our philanthropic activities we aim to strategically align donations and in-kind giving with our business and its interests'.

Board support

'Our board has an absolute commitment ... an assumption that we can't be in business without it. As a listed company we have obligations for this. ... Lend Lease has a multinational board that has different cultures and philosophies. For example, if they come from the US they see it as an obligation. If they come from the UK they see it as compliance. But it is never challenged. We are very lucky we have a board that doesn't need to be convinced'.

External barriers to adopting sustainability

For Lend Lease, the lack of international agreement for the built-environment sector on how sustainability performance is measured is a significant issue – 'We can't currently compare Lend Lease across sectors because we don't have a set of measures that can be used by investors who see the construction and development industry as high risk and not contributing a positive legacy to society. Without those indicators there is a biased view of our sector and it's hard to defend the development industry against accusations and show that we make huge contributions to the social fabric of society perhaps more than any other sector. We do not adequately measure it, benchmark it against best practice in our own and other sectors, and can't demonstrate it as easily as some other sectors can'.

Internal barriers to CSR

One internal barrier to promoting sustainability in Lend Lease is adequate resourcing – 'having enough people to gather the data for the outcomes that we are trying to measure and report'. Another barrier is education and creating a wider understanding across the business (outside the sustainability group) of the contributions that people make to sustainability – 'We have cracked the awareness because we are a sustainable organisation. We have real projects of excellence to help us articulate what needs to be done to achieve it. The next piece is embedding it across the traditional functions. We are re-educating the current workforce. It's a moment in time'.

Developing a sustainability culture

The Global Sustainability Group has promoted the development of the sustainability culture through a global employee awareness campaign – 'which was about unlocking what was already there and celebrating success and providing education opportunities for people to learn about it'.

A dedicated sustainability website (www.lendlease.com/sustainability) has also played an important role in defining what sustainability means to Lend Lease. The website states, 'While we recognise that there is no one definition of sustainability applied to or across industries, we have worked to build a common understanding among our stakeholders of what sustainability means at Lend Lease. ... Broadly, we believe sustainability encompasses how we conduct our business, now and in the future, through the pursuit of workplace safety, a commitment to corporate social responsibility, environmentally sustainable solutions and employee diversity, development and opportunity'.

There has also been an attempt to ensure that employees know that sustainability is valued, reported and measured – 'We have tried to ensure that it's part of peoples' job descriptions since everyone has a role to play. Often it's not until it's in a job description that people engage'.

The company identifies its culture by the voluntary way in which employees advocate sustainability issues – 'Employees will raise sustainability issues around a project whether a client wanted it or not. But having a culture is not enough. For example, hearts and mind were completely given over to safety yet we were still experiencing fatalities. We had the culture but we needed the systems and processes to deliver the outcomes. ... It's the same in sustainability. Projects of excellence now naturally occur because the culture drives that innovation'.

It is notable that the Lend Lease culture in Australia is very much about sustainability not CSR specifically. The reason for this is that – 'In the UK CSR is accepted, our UK business produced CSR reports, but in Australia it's sustainability and if we started to talk about CSR here there would be confusion about it. We see sustainability in a broader context and social sustainability fits into it, not the other way around'.

The main challenges in successfully implementing CSR strategy

'Our main failure has been our failure to realise that this process takes time'. Systems and processes are seen as the main challenge in ensuring consistency of approach across the whole business – 'The global financial crisis gives us the opportunity to focus on efficiency and refocus our systems and processes getting ready for the return of the market. ... It's wonderful to have that internal focus and if we fail, it's because we didn't take advantage of this opportunity to align our internal systems and processes'.

Education is also a challenge as are defining roles and responsibilities for sustainability – 'We would have brought in the education for senior managers much earlier. I also wouldn't create so many people with sustainability in their title. I would create the understanding that sustainability is everyone's responsibility. I would have brought in earlier, the roles and responsibilities of each discipline.

We have had KPIs for people for seven years (70 per cent financial, 30 per cent non-financial) but only now have moved to 50 per cent financial/50 per cent non-financial and we are really interrogating people's roles and responsibilities on the ground.'

Finally, as a global company, a key challenge is to ensure a consistent approach to sustainability across all business units in all regions in which Lend Lease operates – 'We have built a network of Sustainability Executives who are charged with the responsibility of developing, reporting and maintaining sustainability initiatives'. The Lend Lease website states that Sustainability Executives in each business unit provide a strategic framework for knowledge sharing, for advocacy with key stakeholders and for effective communication with the Global Sustainability Group regarding corporate imperatives and requirements. The listed responsibilities of Sustainability Executives are extensive and are key to building a sustainability culture. Among other things, they are required to incorporate sustainability vision, policies and objectives in the development and implementation of business-level sustainability business plans. They report progress of these plans and collaborate with each other to share knowledge, materials and resources that will directly support the mainstreaming of sustainability and green-building knowledge and practice across the regions. They also support the business unit leaders with knowledge and networks, and report to the Global Sustainability Group on matters relating to current and future activities arising from sustainability business plan implementation. Externally, they are responsible for forging partnerships with industry and professional peak bodies, public and private education and research providers, and other organisations relevant to sustainability and they advocate to government, non-government and industry to ensure the removal of barriers to sustainable development.

Measuring ROI

There are clearly perceived investment, capital raising and reputation opportunities which derive from being listed on the Dow Jones Sustainability Index (DJSI) and using other international reporting instruments such as the Global Reporting Initiative (GRI). The Lend Lease website states, 'The value of sustainability and good governance to running a successful business are shown by comparing returns from companies in the Index with the rest of the market. Since 1999, companies on the DJSI World have outperformed the market, in particular over the past two years … According to the most recent (2008/2009) reported review of companies on the DJSI World, Lend Lease achieved the 'best score' in our industry group for our corporate governance, climate change policy and standards for suppliers'.

However, internal ROI appears more difficult to measure, particularly in the social realm – 'Our CEO says we are there – we have arrived – what we have to convert now is the commercialisation of sustainability – demonstrate that there is a ROI. The way of doing this is well established in financial and also now in environmental performance but there is a long way to go in the social performance'. Lend Lease's website states, 'Until now, our philanthropic and social investment activity has been integrated into our business planning minimally. At project level, our businesses take a sophisticated approach to community investment, but there has been no guiding principle that unites the approach from one Business Unit to the next'. To address this, Lend Lease has spent the past year developing social investment and philanthropic principles and reviewing the social and ethical considerations that should guide its investment decisions. Additionally, representatives of its businesses have been working to develop a globally applicable 'tool box' of methods and measures that will help to replicate good practice in projects around the world. The website states, 'We are, for the first time this year, able to report a reasonably accurate global philanthropic cash contribution of AU$3 million. Our Business Units will provide this information from now on as a measure of our philanthropic and social investment. We are yet to quantify the value of time Lend Lease employees volunteer in the community'.

Clearly being a global business, the ROI differs in different regions. For example, in some parts of the world where legislation is advanced, Lend Lease's ROI is that it is out of business if it doesn't embrace sustainability – 'But ultimately to measure the ROI of the global sustainability group is subjective. It's about things like market leverage, opening doors to go and talking to potential clients. We would also say that the ROI is that we comply and that we won't pay huge fines ... In other words we are saving ourselves money by acting now'.

Reporting sustainability

The Lend Lease website says, 'We recognise the importance of maintaining investor confidence through full and timely disclosure to our shareholders and the market of relevant and accurate information about our activities, in line with our External Communications and Continuous Disclosure Policy, which sets out the protocols applicable to directors, executive officers and employees'. While Lend Lease sees measuring, accountability and reporting as key to sustainability success, its main progress has been in improving financial and environmental reporting rather than in reporting social sustainability outcomes – 'We haven't yet found a way of reliably measuring those outcomes. This is part B of the strategy that we are working on at the moment which involves working in our own industry and with the GRI to develop a sector supplement that allows us to benchmark ourselves across the industry and recognise the contribution we make as an industry to society. If we can't measure and quantify it, it continues to be undervalued. In environmental measurement and reporting there is widespread agreement but there are virtually none on social reporting to allow us to know if we are ahead or behind the sector. There are no internationally agreed indicators for social impacts. Occupational health and safety (OHS) is an exception – we now have a way to measure OHS data across all our projects globally and we have a dashboard that allows us to monitor our performance in real time and benchmark it'.

Clearly the scale and diversity of the business presents significant challenges for reporting – 'Lend Lease is the only company in its DJSI sector classification (Real Estate Holding and Development, Financial Services) to operate on a global scale. This makes our non-financial reporting more challenging, given the unique social and environmental priorities in each of the more than 40 countries in which we operate ... Benchmarking our performance against a range of social and environmental indicators is critical to achieving our sustainability aspirations'. The reporting indicators used by Lend Lease include GRI, Dow Jones Sustainability World Index (DJSI World), Carbon Disclosure Project, etc. In addition, Lend Lease also participates in a number of other frameworks. These include assessments of corporate non-financial performance by analysts such as SIRIS and OEKOM who research company performance on behalf of investors and subscribers.

In reporting its performance, 'We try to be as open as we can. For example, we were very open about our poor safety performance. We failed horribly because we had nine fatalities in the last year throughout the world. But in reporting that as a failure, we also make the commitment to learn from these incidents and understand how to prevent this in the future through better reporting and data collection, etc. Also our CEO will say openly in our carbon disclosure report that we haven't done as well as we wanted in reducing our emissions but we are working on it. All of our environmental data is on our website and it shows very clearly what we do. A lot of our data is voluntarily reported. For example in the USA there is no requirement to report waste and recycling but we are now collecting and reporting waste data for projects across the USA'.

Further information about what is exactly reported can be found in Lend Lease's online sustainability performance report (www.lendlease.com/sustainability), within the External Communications and Continuous Disclosure Policy and the Corporate Governance section of the Lend Lease website. In summary, environmental reporting currently includes energy and materials consumption, waste generation, environmental condition of land prior to development activity, land space occupied by asset and use, transportation impacts, water consumption and ecological footprint. Employee data reported includes: OHS statistics; employee demographics such as gender, age, status and length of service; employee turnover; diversity; parental leave; stock owned by employees; and number of graduates; etc. Community data includes social investment and philanthropic activity.

Key determinants of sustainability success

'A third is advocating the change which is bigger and broader than your business. That's true leadership which forces you to engage with thought leaders internationally. A third is celebrating the success that at first may seem random and one-off. But really run at it. Don't be frightened. Leverage it and market it. It's absolutely about advocating your position. The final third is to develop the systems and processes to back your claims up and reporting and feedback loops'.

Stockland

Stockland was founded in 1952 and was listed on the Australian Stock Exchange (ASX) in 1957. Stockland is one of Australia's largest, most diversified property groups with assets at 30 June 2008 of over $14.7 billion in Australia, UK and Europe.

CSR drivers

Stockland's Corporate Responsibility and Sustainability (CR&S) strategy is set out in Stockland's annual CR&S Report where the drivers are stated as balancing responsibilities to Stockland's stakeholders (employees, customers, investors, government, authorities and communities) and its impact on the environment. Descriptors used in communicating the CR&S strategy include: 'doing the right thing'; 'integrating social, economic and environmental performance'; 'future proofing the company'; 'reputation'; and helping the company to achieve its 'business objectives'. The stated focus of CR&S in Stockland is to help it 'understand and have positive and ongoing relationships with a whole range of key stakeholders', to realise potential business benefits from operational savings in areas such as increased energy efficiency and be able to influence regulations and government policies – 'We saw a direct link between CSR and the performance of our business.'

CSR strategy

Stockland's CR&S strategy is guided by 'a continually shifting understanding of our risks and opportunities ... it guides how we work with our stakeholders and has definitely evolved over time. It's not to make a big song and dance about it because we open ourselves to potential criticism and we can avoid accusations of green wash. We just get on with it and make the business case for our customers and how it can benefit them'. Stockland's strategy draws from the assurance standard AA 1000 which focuses on key stakeholder needs, identifying their material concerns, demonstrating responsiveness to these concerns and how this is integrated into business strategy and reporting. Stockland's annual report states that the strategy is reviewed every twelve months through a series of steps, which begins with its business goals and proceeds to collecting key stakeholder material issues, and prioritising them, and then to discovering concerns about Stockland's performance. Stockland's strategy is about trying to align these interests which is then taken to the Board CR&S subcommittee for sign-off.

Who was involved in developing the strategy?

The annual report states that Stockland has established a CR&S Board Committee which is chaired by the deputy chairman with the chairman and MD as members. It meets four times a year and receives a monthly report on CSR performance against industry benchmarks – 'It is fair to say that the board has driven the CSR strategy from the top. Four to five years ago a number of key things came into play which caused us to think more strategically about CSR. In particular we were interested in other organisations such as Westpac which had used CSR to turn around customer dissatisfaction. And we also had an executive from Westpac come over to Stockland who was quite critical and our MD was also very interested at that time … Managers also have a role to play in the formulation of our CSR strategy through specialist sustainability managers across the organisation. CSR was also driven from the grassroots bottom-up as we expanded rapidly with many new employees asking questions about what Stockland was doing in this area'. The main mechanism for this grassroots influence is the CS&R Employee Committee which has the responsibility to shape and annually refresh CS&R strategy and develop action plans which prioritise the most important issues. This committee meets monthly to review progress against the plan and holds a two-day workshop each year to help prioritise the most important issues. 'We also consulted our major clients, customers, business partners and communities to discover how we can work more effectively with them. More than 50 per cent of our projects now have a community consultation plan; we have sponsored an international study on housing affordability; we have established a sponsorship, volunteer, giving and community donations strategy to connect our business to the communities in which we work; set community development guidelines for all projects; and at our strategy day we invited key stakeholders (including climate change advisors, environmental investors and tenants) to give us feedback on how we have gone over the last year and where we could work more effectively in the future'.

Finally, voluntary involvement in numerous external organisations also appears to have influenced Stockland's CR&S strategy. For example, the annual report claims that Stockland are working towards establishing green partnerships with tenants to increase sustainability outcomes in its buildings and have partnered organisations to run energy forums for their tenants to ensure they behave more sustainably. Stockland also states that it is an active supporter of and contributor to the Property Council of Australia and the Green Building Council of Australia and has also contributed to the development of the NABERS energy and water efficiency tool for shopping centres.

Board support for CSR

'It's been a journey for all of us. We have all learnt along the way. They (the directors) now go and talk about what we have done at Stockland to major forums such as the Institute of Company Directors and if they sit on the board of other companies our experiences are transferred there … Do they believe in CSR? I am always very careful about using the word 'belief' to avoid it sounding like a religion. I must always talk about the benefits to the success of the organisation and why key stakeholders expect us to respond to key issues which then drives the business case as to why the business needs to change to respond to these issues. CSR strategy must be built on evidence as far as possible and be kept objective – using as much evidence as we possibly can from key stakeholder groups and keeping emotion out of it'.

Barriers to CSR

Perceived external barriers include confusion about what CSR actually is, scepticism about what sustainability is and what it stands for and how these are addressed. There is a perception that there is a lot of greenwash in the industry and that this has caused Stockland to be very careful about overselling its record and of then being accused of being inaccurate in its claims. There are also considerable regulatory barriers because companies such as Stockland have to cut across state and local government restrictions which means that delivering on any innovation can be risky and costly – 'There are few incentives and rewards and so there is often a penalty for pioneers'.

Internal barriers include sceptics – 'While we will always have sceptics (for example people that don't believe in climate change) we make it clear that we are not here to change personal values but simply help the business succeed long-term and that we therefore need to listen to what our stakeholders think and what they are concerned about.'

There are also difficulties in quickly and easily measuring ROI – 'so we have had to come up with our own metrics'. Stockland's annual report shows that it uses a range of established metrics such as NABERS and Green Star, London Benchmarking Group, and those provided by GRI and UN Principles for Responsible Investment, etc. However, it is clear that there are different challenges with different parts of the business – 'For people in the field, the requirements are clear-cut since they are set by regulations but it becomes harder for other people such as finance managers because the impact of CSR on their job is less easy to define and measure. It's particularly difficult when talking about social sustainability and impacts of communities'.

Another barrier is that CSR is complex, so Stockland seeks to give employees simple measures and tools to use. They are also encouraged to keep initiatives small at first without becoming too ambitious. Stockland encourages employees to do this by linking CSR objectives into performance evaluations for different job families – 'For example it may be about how much they contributed to the culture of sustainability or engaged in our volunteering programmes and other initiatives that we run'.

A final identified barrier is integrating CSR into Stockland's supply chain. In 2007 the annual CR&S report points to a 'sustainable supply chain management programme' and a 'supplier code of conduct' to ensure there is uniform reporting of CSR performance which in theory, better integrates and selects suppliers that understand CSR in the same way.

Developing a CSR culture

The culture is said to be driven from the top – 'We really focus on messages coming from the MD as the person responsible for the performance of the entire organisation rather than the sustainability manager. This shows that we take it very seriously. It's not a bolt-on and that it's part of the business and that the top person takes it seriously'. Stockland also builds CSR into people's personal KPIs and reward systems for employees – 'We have objectives for all job families and we have sustainability managers embedded in all key business units so we don't have a large bolt-on sustainability team but sustainability managers throughout the organisation to affect change and build CSR into the business strategy of each business unit'.

Challenges in successfully implementing CSR strategy

Getting senior managers to engage with longer-term type strategy when many drivers are short-term is problematic – 'The current challenge is with the financial crisis which means getting the head space of senior management is difficult ... Their commitment and our resources have not changed but the time they can give to it has'.

Another challenge is being able to quantify the success of CSR strategy in hard dollar terms – 'People often want to boil it down to dollars but this is not always straightforward because you are projecting out a couple of years into the future or about the dollar benefits of better engagement. So you need to get smart about communicating this'.

Measuring ROI

There is a deliberate, if not always straightforward, attempt to measure ROI in Stockland – 'We set annual targets such as energy consumption and carbon emissions (because investors are increasingly interested in this) and measure against the whole portfolio so that we can raise the bar entirely, not just on individual assets'. Stockland focuses on a five-year-payback period, particularly on fortress assets which they wish to retain into the long-term. But measures of ROI are not always confined to dollar outcomes – 'For us it boils down to enhanced reputation and employee engagement which we try to measure through customer and employee surveys'. Stockland also subscribes to various indexes such as the London Benchmarking Group index to measure community contributions in cash and in-kind through initiatives such as donations, volunteering programmes, supporting local sports teams, etc. – 'and compare ourselves year on year as a percentage of profit'. Another ROI measure which Stockland's investors are exploring is lease expiry with government tenants who are forced to locate in buildings with certain environmental standards. Finally, Stockland employ other metrics relating to areas such as OHS, levels of influence on government policies, stakeholder engagement, supply chain engagement and partnering to help top-tier partners to manage risk.

It appears that one of the biggest problems with measuring ROI from CSR is – 'that people expect you to get it right and the benefits are not always evident when you do. But when you get it wrong you know about it'. The point being made here is that the real ROI from CSR is not just the benefits but the unquantifiable savings obtained from avoiding a problem which can be significant but never known – 'the difference between getting it right and wrong'.

Reporting CSR

In 2006 Stockland published its first CR&S report, a companion to its annual report, setting out its plans for CSR, sustainability, documenting progress to date and giving an account of its environmental, social, governance and financial performance for the previous year. The report applies the AA 1000 principles and reports on metrics which include employee engagement, asset GHG emissions, asset water and energy rating, stakeholder engagement, corporate governance, climate change, labour turnover, ethnicity, gender, training and development, OHS, supplier engagement, customer satisfaction and community support. Stockland have also self-reported against GRI criteria (Australia only) which include economic, environmental, labour practices, human rights, and social and product responsibility criteria. Stockland also report against the UN Principles for Responsible investment.

Since Stockland subscribes to AA 1000, it is required to report where it did not deliver against its targets. Stockland acknowledges that this is not always easy – 'We have had some failures and this is reported openly in our annual CR&S Report. Here we report the bad news as well as the good news and the lessons we have learnt from our failures and how we will put them right. For example, the Sandon Point project was very controversial and there was a lot of media flack that we copped ... A high performance organisation will generally find it difficult to talk about failures and lessons learnt since people just want to get on with the future. But we will do the same in our next report where we will talk about another challenging project and how we will remedy and improve the relationships with the community. You have to be open and honest and being open to failure will ultimately lead to greater trust with stakeholders but it takes a while for an organisation to get there and be prepared to go into that space'.

Key determinants of CSR success

Making sure the board is committed is the most important determinant of CSR success for Stockland – 'Having a supportive board is critical and having a board subcommittee has been a very powerful vehicle which sends a strong message that CSR is important ... The role of the MD and the CEO is also critical. They must talk about CSR, rather than the CSR manager, and when we have a quote for the media it should also be from them'.

Establishing a common understanding of CSR is also identified as important – 'People throughout the business should also be able to talk about what CSR means to them, to Stockland and how they contribute to that so that everyone feels part of the journey and success'. Finally, CSR must not be a bolt-on but be fully integrated into all parts of a business and metrics have to be developed to measure performance – 'You have to have support from business unit leaders, they have to see what success means for them and then they will build it into their individual business unit strategies'.

The GPT Group

GPT has been listed on the Australian Securities Exchange (ASX) since 1971, and is one of Australia's largest, diversified and listed, property groups with assets of over $13 billion in the retail, office and industrial/business park sectors. The Group is one of the top 100 ASX stocks by market capitalisation.

Drivers of CSR

GPT was originally established as a business unit with responsibility to raise equity from the public for the construction pipeline of Lend Lease. While different parts of the business were experimenting with CSR, it was not until GPT became an independent entity that it was managed systematically – 'It wasn't until 2005 that we were able to decide what we really stood for in this space and then proceed to create that'. For GPT, the language of CSR is about 'doing the right thing' and there also appear to be strong internal CSR drivers which derive from its heritage and history – '... the original concept of the company had some grounding in social equity since it was set up to give people in the community access to a secure property investment which they otherwise wouldn't have ... Social equity has always been a part of the culture of the organisation'. Other claimed internal drivers are employee attitudes – 'our people are generally socially minded and we have a selection process to ensure that our people are motivated to be responsible, conscious of doing the right thing and to do a job that has some values attached to it'.

Externally, the drivers are numerous and include, '... reputation, our peer group of companies in our competitive space which we compare ourselves with, institutional investors who are demanding more responsibility and responding to external pressures themselves, markets which are driving firms down this path, then there are our tenants and the relative position of your assets, clients and business and government partners who won't even let us get to the first base or get shortlisted without some credibility, and there are the communities in which we do business which are becoming smarter'. The unique aspect of real estate from a CSR perspective is seen to be its strong physical manifestation in the community, unlike other infrastructure developments such as mines which may be hidden underground or in remote locations. GPT's products and services are perceived to have, 'strong community drivers around it and this forces you to be very clear and vocal about what you stand for in the CSR space. We must make sure that we deliver on the public commitments we make and being transparent around this'.

CSR strategy

GPT's CSR strategy is to 'move beyond that dictated by its core business of delivering property assets which deliver the best possible rent and highest possible occupancy ... to be a leader in our sector, being best in class and being a responsible company'. GPT require all new developments to be 'leading edge and break ground so over time when we do our 20–30 year scenarios they do not become liabilities'. In the social sphere, GPT's aim is to 'develop its people's skills and integrate that into day to day jobs'. From a community perspective it is about 'integrating our business into the community through partnerships with community groups, employee volunteer programmes, etc. rather than signing cheques ... It's about us being clear about our CSR goals and delivering on what we promise and demonstrating that we have done so'.

GPT strategy in social and community investment seeks to tackle issues in numerous areas such as employability, building community social capital, reducing humanity's ecological footprint, environment and indigenous affairs. The 2008 GPT annual corporate responsibility report states that the long-lived nature of its property investments means that GPT is a business focused on 'inter-generational interests'. The report acknowledges that the sector is also a significant generator of environmental impacts and resource consumption. In response it states that GPT is committed to operating in a manner consistent with 'best practice' corporate responsibility principles and performance. Although best practice is not defined, the report asserts that this commitment involves investigating and pursuing pathways to achieve parity between environmental, social and financial returns. It also states that GPT is committed to being open and transparent in the way it communicates with stakeholders and to this end has adopted the GRI to guide its CSR reporting. The annual report points out that GPT's five-year-rolling strategic plans set out environmental, social and governance goals and GRI reporting is used to verify the performance achievements against these goals. It also states that in developing GPT's corporate responsibility strategy, the Board and senior management have considered the organisation's environmental, social and economic impacts and have compared them with global and Australian standards in order to identify those issues which were essential to respond to in the medium-term – these are incorporated into GPT's 2012 goals. Longer-term issues are integrated into GPT's 2020 goals.

GPT's CSR strategy is said to be driven directly out of 'aligning business and community interests'. GPT states that it is in its interests to have healthy and viable communities and has sponsored research to investigate what social aspects of community are most important to its business objectives and to then have invested in this. For example, its programme to reduce Aboriginal crime in the Northern Territory not only benefits the community but also directly benefits its business by reducing crime in its many shopping centres, making them a safer place to work and shop.

Who was involved in the development of the strategy?

GPT's approach to corporate responsibility is overseen by the board. Reporting directly to the board is the Board Corporate Responsibility Committee to which report the Corporate Responsibility Steering Group (CRSG) and the Corporate Responsibility Working Group. The GPT annual report describes the responsibilities of the various committees as follows.

The Board Corporate Responsibility Committee consists of three independent non-executive directors, a chair and two directors. The Committee has been established to: review the strategic direction set by the GPT board and ensure that it is followed; review the CRSG business plans to execute the corporate-responsibility strategy; review the quality and reliability of the non-financial corporate-responsibility reporting processes; review and report on non-financial corporate-responsibility statements issued by GPT or ratings submissions made by GPT; receive, analyse and assess corporate-responsibility risk and compliance reports; and oversee the risk management, compliance and internal control framework in the context of corporate-responsibility initiatives.

The CRSG is chaired by GPT's CEO, with senior management representation from each area of the business. This Group is responsible for the implementation of the strategic framework, identification of performance goals and their implementation across the business. It also sets the corporate-responsibility strategy and policy for the Group that all its divisions will tailor to their specific needs; guiding risk-assessment framework development, including coverage and monitoring of material external issues; and individuals act as champions on corporate responsibility issues within GPT and externally.

The Corporate Responsibility Working Group is responsible for day-to-day implementation of corporate responsibility in the business. A key goal of this Group is to share knowledge and build capabilities through the business. This group meets quarterly and comprises all sustainability and corporate-responsibility professionals in the business and is chaired by the Head of Corporate Responsibility. The objective of this Group is to share information, maximise collaboration across portfolios, provide mentor support, improve knowledge, monitor compliance and celebrate success.

Externally, GPT is a signatory and member of various CSR-related organisations which will have influenced its CSR strategy. These include the UN's Principles of Responsible Investment (PRI), the Carbon Disclosure Project (CDP), CitySwitch Green Office and Total Environment Centre's Existing Buildings Project, etc.

Barriers to CSR

The main barrier is seen to be fear of the unknown and fear of failure – 'You have to convince people, prove to them that new things work, overcome their prejudices. It's about getting over the barrier ... It's mentally different, getting people to take a risk. People don't know how to start thinking about it and where to start doing it ... CSR is about teaching people new ideas and concepts, helping people to start thinking about it, giving them simple instructions to take away to follow and they then start coming back and helping us improve our processes ... You have to help people through a thinking process that is different to what they have done before ... In our CSR team our skills are not technical. It's about teaching, tenacity, motivation, persuasiveness, bringing about change, passion and getting people to own it'. CSR development is also about having a strong business plan – 'CSR was initially seen as a cost but now we have a proper business rationale ... You have to take the emotion out of it, you have to be able to prove it works for the business'. There are clearly considerable challenges in demonstrating that each initiative goes to the bottom line – especially when talking about initiatives like community programmes. There appears to be little data to verify this although it is claimed that in the current financial crisis there has been 'no push back' from shareholders about CSR initiatives – '... there is an increased appreciation there of its value to the business. Some of our funds guys who have been a bit sceptical in the past are now saying it's real and that we have to do something about it'.

Developing a CSR culture

The main drivers of CSR culture were identified as 'having a strategy, communicating it effectively, having people's performance tied to it and working with every job to ensure there are clear CSR KPIs associated with that job which it can be evaluated against'. Educational and motivational strategies are crucial and involve each job having CSR goals attached to it and encouraging people to achieve them through GPT's Short Term Incentive Scheme. This scheme requires each employee to undertake and contribute to defined corporate-responsibility activities. There is also an internal newsletter that is circulated every quarter, which features an update on CSR achievements across the business, the Group's social investment platform, and a personalised column from volunteers outlining their experience in volunteering projects, etc. – 'Every employee has a volunteer menu which we help to create the necessary connections for'. Every quarter, GPT also runs 'CR Lunch and Learn' seminars which are informative presentations on CR-related topics. The topics vary, and include case studies on some of GPT's environmental assets, information for staff on how to be more sustainable in their homes and workplaces, and talks by leaders in the CSR field. On individual projects GPT also uses footprint ambassadors and ESD (ecologically sustainably development) catalysts. Employees take the lessons learnt from their experience and build upon them to improve future GPT projects, target greater ecological footprint reductions and continue to advance GPT's efforts to educate new and existing tenants. In addition to this, GPT's learning platform and approach to skills development includes corporate responsibility. The corporate-responsibility team works with HR to build a set of competencies and skills for each role in the business. Finally, GPT attempts to integrate CSR into their supply chain by working with the Property Council of Australia to develop its 'Principles for Fair Contracting'. GPT references the Principles in its integrated service contracts for operational assets and requires contractors to abide by these principles.

Challenges in successfully implementing CSR strategy

The main challenge is simple communication both informally and formally – '... You have to package CSR very simply and understand the different channels through which it is best delivered ... Getting the right channels at the right time. But you can do all the formal channels and yet people still may not get it so there is a lot of elevator speak'. It is also important to know where to set the performance bar which can be difficult in an international business because 'there are many different settings ... what does best in the world mean? Culturally it becomes very difficult internationally because people in different countries understand different standards'.

Measuring ROI

GPT has established clear measures of ROI for its environmental initiatives but is still developing ways to measure ROI associated with its social initiatives – 'We are working on developing a social return on investment methodology (SROI) which is a way of monetising things that are normally not quantifiable'. For example, GPT has an indigenous employment programme for which it may measure resources committed (cash, time and in kind), improved health outcomes, employment outcomes, tax benefits to the government, reduction in government benefits, etc. – 'We don't see the cash flow to us but when you are able to change some of the disadvantaged in the community, which reduces crime etc., it does of course benefit our shopping centres. This is core to our business. So any disadvantaged group falls into our basket and we try to help these people to shift established indicators of disadvantage in the community through community volunteer programmes and sponsorship, and promotion of environmental initiatives like the Banskia Foundation'.

GPT has produced an annual CSR report since 2001 (called the sustainability report until 2004). This report adopts the GRI as its reporting framework, and also participates in relevant surveys and benchmarks in relation to its business. Other ratings and measures of GPT's CSR performance include the Dow Jones Sustainability Index which assesses GPT's ability to manage risk and leverage opportunities across the economic, social and environmental agenda. GPT is also registered with the FTSE4Good Index Series which has been designed to measure the performance of companies that meet globally recognised corporate-responsibility standards, and to facilitate investment in those companies. GPT also subscribes to Governance Metrics International's Global Governance Ratings (which assesses a company's corporate governance practices and looks at value-drivers not revealed through financial analysis) and The Corporate Responsibility Index which looks at how sustainability is integrated into GPT's strategy, operational management and reporting.

Outside these formal reporting mechanisms and indexes, GPT also uses well established ESD tools for its office and industrial/business park assets. This includes Green Star which is a national, voluntary, environmental rating scheme that evaluates the environmental design and achievements of new buildings. GPT also uses NABERS which is a performance-based rating system for existing buildings and the One-2-Five® Water tool to help establish effective processes for managing water. In seeking to measure and reduce the ecological impact of its activities, GPT has developed the Ecological Footprint Calculator to help GPT and its tenants understand impacts and take steps to reduce the environmental impact in design and operations. Finally, in other areas such as employee engagement, GPT uses the London Benchmarking Group (LBG) methodology which measures the inputs, outputs and impacts over time of a company's community contributions to benchmark how companies are contributing to the community in comparison with peer companies.

In reporting CSR failures, GPT claims to be as open as it can although acknowledges the care that needs to be taken in doing so and that words need to be chosen carefully – 'There are some things we show we have not achieved on our website and annual report'.

Key determinants of CSR success

The key determinant of success is identified as GPT's people and having a supportive board that people can 'go to with ideas'. It's also about clarifying their objectives and developing measures of ROI. In the environmental sphere measuring ROI appears to be very clear but it is not so clear in the social sphere – '... so you have to pick the ones that have an ability to be measured like unemployment, crime rates, school dropout rates, etc. These are things you can measure and put a value on for the business. Unless you start somewhere with a measure you can't go anywhere. So it's about being clear about what initiatives meet our business objectives. So that when you are approached by a cause, people know whether to say yes or no'. GPT also points to the need to 'slow things down' to the pace that an organisation can work at in terms of its business culture and resources. And there is also the need to develop the 'internal skills to drive CSR throughout a business' – education and training are clearly important factors in ensuring CSR success. Finally, success depends on balancing CSR with core business objectives – '... the paradox of CSR is that you are open about what you do but at the same time open yourself up to more criticism and your competitors. So success in this sense is about balance – creating the value of our company versus achieving a wider social good. At the end of the day if we can connect our people to what GPT wants to be and create a structure which empowers people at the coalface in interaction with the community to come up and give us ideas ... That alone will constantly keep you evolving and responding to local and global issues'.

Leighton Contractors

Leighton Contractors Pty Ltd is one of Australia's largest construction and resource services groups and a subsidiary of Leighton Holdings Limited, a company listed on the Australian Stock Exchange, with around 37,000 employees and its head office in Sydney. Founded in Victoria in 1949, Leighton Contractors has grown from a small, privately owned civil engineering firm to become a large diversified group, with interests in telecommunications, infrastructure investment, facility management and energy.

Drivers of CSR

Leighton Contractors considers sustainability decision-making to be key to corporate longevity, consistent profitability and responsible corporate performance. There are a range of drivers both internal and external. Externally, a growing trend in the markets is that clients are becoming more aware, sophisticated and demanding about sustainability on their projects. There is also a perceived need to respond to changing social values in the community. New legislation relating to carbon emissions is important because Leighton is the largest contract miner in Australia and one of the largest infrastructure contractors, with consequential high diesel and energy usage. Industry tools, such as those produced by the Green Building Council of Australia and the Australian Green Infrastructure Council, are also important drivers of change because they provide a common platform for clients to measure the sustainability performance of projects.

Internally, sustainability is driven by the behaviours of senior management and, as consistently shown in the company's annual employee opinion surveys, there is widespread belief in the company's values and in their role as an integrating framework for achieving shareholder value – 'Three years ago the company was on a journey to define its values. What we have now been able to demonstrate is that if the company "lives" those values it is very consistent with sustainability ... These values are really the drivers of our strategy'. Leighton Contractors' five values are 'safety and health above all else, respect for the community and environment, enduring business relationships, people are the foundation of our success, and achievement through teamwork'.

At a corporate level Leighton has developed KPIs around each of these values. These run down through each of the divisions at an operational level – 'They influence our actions and behaviours and have become drivers of value in the business. So we believe that living our values leads to a sustainable organisation'; and provides insights into the level of traction and effectiveness being achieved at an operational level.

Describe your CSR strategy

Leighton's has an integrated sustainability strategy – 'We have a very diverse business and CSR is like a commonality across the business which fits in with our strategy'. Its strategy is documented in its annual sustainability report and starts with its definition of sustainability. This states that, 'Our sustainability strategy is founded on living our values – through what we do and how we do it – so that we can achieve our core purpose of transforming ideas to enhance people's everyday lives'.

There are four streams to Leighton's sustainability strategy. The first stream is establishing a culture of sustainability through embedding its values in people's actions and behaviours that carry into the day-to-day operation of the business. This is achieved through recruitment, induction, training, personal development performance and development appraisals, etc. The second stream is to integrate sustainability in the business through consistent systems and procedures, so changing the way people work and make decisions, and leading to best practice and continuous improvement. The third stream is strategic planning, which is about using sustainability to generate new business opportunities, drive innovation and encourage continuous business improvement – 'Given the current focus on new legislation, we have had a strong focus on carbon. Looking at this strategically has contributed to a restructure of the company to establish an Industrial and Energy Division which is largely focusing on renewable energy and less carbon-intensive working methods'. The fourth stream is about how Leighton relates better to its clients and suppliers and develops strong strategic alliances with these and other stakeholders to drive sustainability and business resilience. For example, Leighton Contractors has a 30-year-long partnership with Caterpillar, which includes working together to highlight improvements in plant design that will provide mutual benefits.

The overall strategy is underpinned by a knowledge-exchange strategy – 'We share knowledge because being a very diverse company with a large geographical spread means that there are a lot of innovative things happening which are not necessarily communicated beyond the project team. So we are working to strengthen the way we share knowledge so that we are innovating off a high platform all the time, not from a low base'.

Who was involved in the development of the strategy?

The strategy has been driven very much by the managing director who was very focused on sustainability and by his executive management team who appointed a Group Sustainability Manager to report to the Executive General Manager of HR and Sustainability – 'We see sustainability as essentially a people issue which is why HR and sustainability are managed out of one department and work very closely together. We haven't got a separate sustainability group in the company; rather the Group Sustainability Manager's role is an integration role across those responsible for carrying each of the Company's values'. Leighton Contractors' 2008 annual sustainability report points out that in 2007/2008 the board created an Ethics and Corporate Governance Committee and a Safety, Health and Environment Committee to manage performance in these areas. In addition, there are company-wide (and business) networks that meet to discuss improvements, initiatives, policies and standards for safety, environment, human resources, and community and business relationships.

The community feeds into Leighton's strategy at two levels. At the operational level, Leighton has strong links into the community on each project – '... we have pioneered community relations as a profession in our infrastructure division to strengthen this'. At the organisational level, Leighton is finalising a review of its community investment strategy to determine where its philanthropic contributions go – '... moving away from straight philanthropy to more strategic community investment'.

Board support

Leighton's board is highly supportive of the Executive Management Team's approach. There is agreement that the adopted approach to sustainability and CSR is tangible and robust. It is seen as supportive of profitability and the long-term health of the business, while adding value for clients and providing a meaningful place for people to work. This attitude has driven CSR from the start – 'The drivers of sustainability are all self-evidently supportive of profitability in the business and value for our clients. It's just a given. So the strategy was created out of our values, rather than as something that would add value in itself. How can you argue, for example, that safety, looking after your people and business relationships are not drivers of profit? We know that if we are doing our values well then we will have a successful business'.

Barriers to adopting CSR

External barriers were perceived to be minimal and reflective of the level of knowledge, sophistication, and business needs of clients and stakeholders – 'My observation is that clients do want it. What they sometimes don't know is what it means for their business and how it can add value for them. Part of our job is to explain that to them and to bring them up to speed on the latest practices. Five years ago clients may have been a barrier but not now'. As well as being a driver of CSR, external tools were also seen as a potential barrier – 'The danger is that they adopt a standard framework developed by the Green Building Council of Australia or something similar and don't understand what it means and how to interpret it, or what its limitations are. On the other hand, such frameworks, which include those of the Australian Green Infrastructure Council and the Australian Minerals Industry, provide industry leadership to stimulate organisational thinking on sustainability'.

Internal barriers to adopting CSR mainly relate to education – 'Getting a consistent understanding across 9,500 people, with the turnover that entails, is an ongoing process. In addition, we have a responsibility to educate subcontractors who work for other companies, and may not have the same cultural commitment to our company and its values. That said, most people want to do the right thing, and so the challenge for us is how to get a consistent understanding across this large, diverse and geographically dispersed group of how we want to do it'.

To achieve this, Leighton has started at the top. It is seen as the senior leadership's responsibility to drive sustainability into their businesses. Leighton is also developing a 'train the trainer' process to close the gap in understanding and knowledge that its annual sustainability report acknowledges has emerged between senior managers and staff. It is intended that the company culture (theme one of the strategy) will then drive the integration (theme two of the strategy) and the integration will drive the culture, etc. – 'We are doing a lot of good stuff across the business but it's not called sustainability or CSR ... it's about bringing this all together, making people aware that we are doing it and then to doing it consistently. It shouldn't necessarily be anything new. But it's how you tell it as a sustainability story and develop a CSR culture'.

Developing a CSR culture

CSR culture is identified as a measurable aspect of the business, a characteristic which is important to nurture – 'By and large the company has a caring culture. That might be at odds with where the company was ten years ago, but led by the current MD it's a very strong culture made explicit by our values. You know you have a culture because your performance, against each of those values, when you measure it, is good and because people want to come to work with you, you have low levels of disputes and you have good long-term business relationships. These outcomes are all easily measurable'.

The culture is also clearly driven by the firm's values, rather than CSR specifically, and ingraining those values throughout the business – 'There has been a very strong communications strategy around our values in the last two years or so which has come out of HR and communications and been supported by the senior leadership. The link to HR is crucial. Many other companies' sustainability agendas have come out of the environment. But we have been very strong that it's not just about the environment but about the social and economic dimensions as well – it is about being a sustainable company – long-lasting, consistently profitable and corporately responsible. So we have morphed our CSR strategy to be a corporate-responsibility strategy'.

Challenges in successfully implementing CSR strategy

The main challenge is building understanding throughout the organisation and integrating it into the business at all levels – 'It's understanding how sustainability suits your organisation. Why is it important and what does it mean to it? Unless it's real and meaningful it's not going to happen. It's the same as any other major cultural change programme. You can't just turn the Titanic. That's why we have integration as a core part of our strategy'.

Measuring ROI

ROI associated with CSR activities is not measured formally – 'It's seen as integral to everything we do so it's not separated out and measured as a specific thing. Of course any investment has to be justified but everything is driven out of our values so we have KPIs for each of them to measure performance'. The environmental indicators are well understood in Leighton and are particularly well embedded in the construction and resources businesses because there are so many compliance issues. Since all competitors have to be similarly engaged it's not regarded as a significant opportunity for competitive advantage, although 'doing' compliance well can give advantage. Externally, the social sustainability indicators are 'not as well thought out' as the industry is less sure of what 'social' means to sustainability performance. However, Leighton has defined sustainability more broadly than just the environment which is perceived to be in contrast to many of their competitors – 'We have a good understanding of the social dimensions and see this as a competitive advantage'.

Reporting CSR

Leighton Contracting produces an annual sustainability report which documents its performance, including that of its subsidiaries and all joint ventures – 'It is meant to be an honest and transparent picture of where we are on our sustainability journey including areas where we need to improve'. The annual report shows that Leighton uses the GRI reporting framework (first used in 2008) and reports on KPIs across various categories of performance (some acknowledged as not achieved) including: financial (revenue and work in hand, governance and risk, structure, policies, risk and opportunity management systems); safety (challenges, key initiatives, injuries, breaches, 'lost time frequency' rate, training and audits); environmental (ecological footprint, emissions, use of natural resources, energy use and incidents); respect for the community (community relations, employee awareness, community investment and employee perceptions); business relationships (knowledge sharing and negotiated work with new and existing clients); people (leadership and talent management, employee development, remuneration and benefits, turnover, employee perceptions, communications and satisfaction survey); and teamwork (internal partnerships, teamwork excellence and employee survey on teamwork).

Key determinants of CSR success

The key determinants of success are senior-management support, alignment with values and understanding – 'The managing director and senior management have to buy in – otherwise you are dead in the water ... Any strategy must be aligned to an organisation's values. Senior management must understand what it means for the business. Your senior managers must be comfortable that CSR is aligned to the business and that it's related to what individuals do on a day-to-day basis. It must be clear how it helps people achieve their goals. Education is also a key factor'.

Kell and Rigby

Kell and Rigby (K&R) is one of Australia's oldest family-owned businesses. Established in 1910, K&R is a fourth generation family-owned construction company which has offices in Sydney, Brisbane and Canberra. It has a functional structure which focuses on different market segments and over the last five years has completed over AU$700m of construction projects in commercial, retail, residential, age care and education.

The drivers of CSR

In K&R, CSR is first and foremost identified with the culture and heritage of the company. It is seen as something which family businesses naturally do, which does not need a label, that has always been assumed, never formalised and a core reason for K&R's longevity – 'CSR is a company supporting the community. I see it in a karmic sense to give back at least as much as you take away ... It's not driven by marketing otherwise it would be easy to get cynical about these things'. The K&R culture is claimed to be driven by three core company values. The first is the golden rule 'treat others like you expect to be treated'. The second core value is 'care for the building' and the third is 'keep progressing'. This formalisation of core values which underpin the company's culture is a relatively recent thing. The culture of K&R has historically been undocumented and unspecified and likened by the CEO to an aboriginal tribe – 'Something was passed around, people would talk about it but it was not codified. It's like our values. Our values were not codified for 95 years and only recently have we felt the need to codify them'. The reason for this is that with recent growth, the company is in a state of transition, moving from a family-based culture which was learnt and experienced through many years of loyal service to the company, to a more formalised culture which needs to be taught and internalised more quickly to a more transitionary workforce. In the past, it was not unusual for people to work for K&R for their entire working lives but this is becoming increasingly rare. It is likely that at some point in the future that this may also lead to a more strategic approach to CSR but for the time being it remains a largely ad hoc and un-strategic activity which is not linked directly to business strategy or goals.

K&R acknowledges that there is a link between CSR and business success, although this did not appear to be formalised or to be a primary motivator – 'There is a bit of bottom line in it ... for example, we could just give Best Buddies (a national charity) some money but I would rather not. So if there is a smarter, more strategic way to help them that uses our strengths, that's a better way to do it ... there are so many causes to give it to. But if there is something where your people can benefit out of it, while being good, then that's great. That's the joy of business'.

Although not formalised, CSR has been a long-established criteria in the selection of clients and business partners and in the treatment of employees and subcontractors – 'if you have a subby that is active in his community, has clarity and is genuine about his role then we can establish a healthy and long-term relationship. We have decades long relationships with some such subcontractors'. While there was a view that it would be good for K&R to be more strategic in its CSR activities, there was a strong belief that it should not be marketing-led – 'It is a good idea to get more strategic about it but I certainly won't let marketing run it ... if something is marketing-led then the community will quickly pick up on it. This could completely unravel any financial benefits from it ... it is crucial that we separate CSR from marketing and do it for our own reasons (karma, not neon). In a large company CSR would be run by marketing as a quasi-marketing exercise. In K&R it is run by the business manager and CEO (and maybe HR involvement), with marketing involved in internal communication but no external marketing at all. This is a very subtle but important point'. The historical craft-based values of K&R have relied on its reputation to get business rather than on self-promotion and blatant marketing, and this attitude extends into CSR – 'The glossy, superficial side of marketing has always been a bit of a dirty word in K&R. If we're doing something for marketing's sake then we shouldn't be doing it. If we are doing it because it's educational, it fits with our values, it benefits our people or our clients then chances are it's the right thing to do and the funny thing is that it is the best possible marketing activity we can ever do'.

Describe your CSR strategy

There is no formal CSR strategy in K&R. Rather CSR is seen as something that has been given a new label by big business that family companies take for granted and is central to their longevity – 'It's something that family companies – especially family companies that have been going for a long time – have always done. One of their reasons for surviving so long, is their integration with the community. I think you will find that with a lot of family companies ... that it's been an integral part of the business through links with local footie teams, etc. For example, Grandpa was involved with the Uniting Church, Dad is involved with several not-for-profits and I am involved with surf life-saving. Where we are involved individually, Kell & Rigby gets involved'.

The informality of CSR in K&R is reflected in the fact that a number of activities, which in other companies might be considered CSR, have never been formally categorised in this way. For example, K&R and St George Bank partner to sponsor an annual International Aged Care Scholarship. This is clearly strategic to the K&R business (which builds aged care facilities) but is not formally linked into an overall CSR strategy – 'We are involved in providing age care and I love the irony that the builder who stands to profit from it, saying "guys it's all about community care, let's stop building these things" – it's almost as if the supply of these things is creating the demand and, if that's true, that's a scary thing'.

The informality of K&R's CSR agenda is reflected in the unstructured way in which decisions are made about causes to support – 'If a random club needs a new rescue board and asks us to buy it, we tend to shy away from that. Yet if there is someone who has shown they have raised some money, then we will say "yes". Or if there is someone who has been with us for ten years and they manage their local soccer team we will say "absolutely". We get a request then we throw it around and if it seems to make sense then we will do it'. K&R is also very pragmatic and tends to focus on 'what it is good at'. One example is K&R's involvement in Best Buddies Australia – a programme in which K&R staff help kids with lower intellectual capabilities integrate into schools through mentors and develop leadership skills – 'So we can give more than just dollars and our people can improve from the experience in a karmic sense, but also when you teach, you learn in a different way, and so they learn about leadership in a whole new way'.

Who was involved in the development of the strategy

The strategy is driven by the family – 'It's at the dinner table. You know, Grandpa, Grandma, Dad, Mum. The development of the strategy is directly intertwined with the culture of the company'. The board has not become involved with CSR at all although 'it has always supported me'. The board of K&R for the first 97 years were the senior executive team. The current board is quite young and has one executive, two independent directors and an independent advisor. In the last two years there has been a deliberate attempt to put a new governance structure in place and form a proper board with external directors to monitor and challenge the CEO. The family is represented by the Family Council which meets annually to approve director appointments, discuss long-term plans and monitor progress.

Barriers to adopting CSR

The lack of formal CSR strategy means that there are no identified or definable external and internal barriers. CSR decisions are largely centralised and K&R contributes to the community in an informal and spontaneous manner rather than in any strategic and planned way which is linked to a business plan. The main challenges in successfully implementing a future CSR strategy in K&R are not to get carried away with individual projects, to ring fence it, to avoid scope creep and 'not forget we are builders ... to keep it hard edged and tempered'. Also, being a project-driven company there is also the challenge of 'keeping people motivated when they are very busy'.

Measuring and reporting ROI

K&R does not formally measure ROI of its CSR activities – 'it's done over dinner. It's not an agenda item'. There are a number of reasons for this. First, the lack of formal strategy and objectives means that there is nothing to report against. Second, the strong disassociation of CSR with its marketing strategy prevents K&R from openly promoting or talking about its CSR-type initiatives. Third, K&R has not defined what CSR success means to its business which means that appropriate metrics will not have been identified. Finally, CSR is claimed to be ingrained within the company culture and has never been formally recognised as a separate and distinct activity.

The fact that CSR is not formally reported does not mean that a sense of CSR does not drive decision-making processes in K&R. On the contrary, the CEO pointed to what he described as a 'beautiful irony that those firms that say they don't practice CSR probably practice it the most'. The CEO's view is that CSR has come about because firms have become disconnected from the communities in which they do business. So they have had to invent a new term to help them to reconnect.

CONCENTRIC

CONCENTRIC Asia Pacific is a small- to medium-sized privately owned firm of consulting engineers and designers which is a provider of technology for 3D design, simulation, interactive training, virtual ergonomics and visualisation to the construction and engineering industries and other industries such as car manufacturing, automotive, defence and mining, etc.

CSR drivers

CONCENTRIC's CSR strategy is driven by the personal passion of its MD to see Australia regain its position as a leading engineering nation. The MD also needed a mechanism to help the company grow and wanted it to be perceived to be doing 'great things'. Driven by a personal interest in how children make career-decision choices, the MD undertook doctoral research which directly informed the development of CONCENTRIC's CSR strategy. This research argued that people are genetically wired to seek and follow heroes, peer groups and ideas which are perceived to be 'cool'. It led directly to the establishment of the Re-Engineering Australia (REA) Foundation in 1983 which has since become the foundation of CONCENTRIC's CSR strategy. REA is a not-for-profit public company established to raise the awareness of modern engineering design and manufacturing careers through initiatives targeting young Australians. Now a global initiative utilising heroes and role models to provide social influence to inspire children, REA runs a series of career intervention programmes which are aimed at increasing the self-efficacy of students toward careers involving mathematics, science and engineering. The main programme is the F1inSchools™ annual challenge which is open to all Australian high school students from years 5–12 to design, manufacture and race CO_2 powered F1 style model racing cars. Teams compete in school, regional and state finals and the best Australian Teams win a trip to the F1inSchools™ World Championships in London which brings together teams from 30 countries. Companies and schools are invited to become sponsors of the REA Foundation and benefits include a certificate of membership, entitlement to use the REA Foundation logo and supplied text/material in company presentations, marketing materials, regular REA Foundation newsletters, recognition of membership on the REA Foundation website, opportunities to collaborate through REA Foundation programmes and opportunities to be inspired by the capabilities of young Australian students.

According to the MD, the business rationale behind the REA Foundation is a belief that 'If you are going to be a leader, then you must do what leaders do ... You have to be inspirational, a hero and you have to be cool ... when it comes to building a corporate image, advertising just like everyone else doesn't work ... As a small Australian company you have to be different to compete against the likes of giants like Siemens ... if you do what leaders do, it will get you into the boardrooms. We wanted to make our company a role model, a hero to children who were about to make their career choices ...'. CSR is an important part of CONCENTRIC's marketing strategy which is designed to pull people to the business rather than push them to buy its services or products. CONCENTRIC deliberately avoids overt association with REA but its sales people mention it at the end of all marketing presentations to help differentiate the business. According to the MD 'it has an amazing impact on getting customers to engage with what we are saying. It softens everything up and gives the customer the impression that we are not just here to flog them boxes or get the highest price for our products ... we are a company that cares'.

CSR strategy

The REA Foundation has been the entire thrust of CONCENTRIC's strategy – 'To us it's always been a marketing strategy ... It's subliminal marketing

... It's in the background and it's about building the trust of our clients in a company that deals with services which are critical to people's safety'. The MD also believes that it comes back to the neo-Darwinian theory that informed his research, the mindset of the CEOs, 'who are CONCENTRIC's customers?' and the focus of the marketing strategy – 'All of them have children and when we talk about what we do with children and how we help children achieve great things, it's very easy to gain an empathetic relationship ... People do want to help, but in the business world they have to justify it back into their business and we help them do this'.

The REA Foundation was neither planned, nor was it formally articulated as a CSR strategy – 'we went in with blind determination'. Yet it has grown out of all proportion to its initial goals and become so successful that it was established as an independent company which is run by the MD as chairman. As the MD points out, 'Now it's driving CONCENTRIC rather than the other way around and has a life of its own ... it is now something that other companies are connecting to because they haven't the resources of their own ... something we encourage'. So the REA appears to have become a vehicle by which other firms and stakeholders in CONCENTRIC's supply chain, and indeed important customers, are able to connect into as part of their own CSR strategy. 'For example, the local paint shop can paint the students' cars for free and the Department of Defence see it as an important mechanism by which they can engage with school children and give something back to the community'. Through this process, these organisations then become linked to CONCENTRIC's business and become potential clients and business partners. 'Our strategy has been driven by our belief that people want to buy from people they trust and know. They don't want the hard sell. Especially Generation Y. Senior executives who are reaching the end of their careers and who make purchasing decisions also generally want to give something back to the community. Most have families of their own and have an emotional attachment to children. They can relate to what we are trying to do. Too often the business world stops you from letting the heart come out because you have to be so bottom-line minded. So when something presents them with an avenue and opportunity to give back to children and take part in playing a hero role then their minds are opened. It's a far better way to talk to people than approaching them on business grounds. While there is a time and place for a hard-nosed approach to business, it turns most people off. CSR is a very powerful marketing strategy and we find that most people are just bursting out of their skin to get involved in what we do'.

Who was involved in the development of the strategy?

Initially the CSR strategy was driven by the MD and the board through the creation of the REA. Although the REA now exists as a formal Foundation with its own separate organisational structure, CONCENTRIC's staff help to manage and run it, contributing ideas which take it forward. The MD continues to be the REA's figurehead, often being asked to present at high profile industry and governmental events – 'As an example, I was asked to speak at the Defence and Industry conference in front of 1,500 people alongside the Federal Defence Minister. They asked me to speak because they are so proud to be associated with REA. I accepted only on the condition that I didn't speak and that students would take my place. When young Australians students speak, they are totally engaging and they blow people away with their enthusiasm. These are the types of opportunities which I could not have had without REA. What is actually happening is our clients then help to promote our business which then draws in other people who want to be associated with it. CSR breaks down barriers'.

Board support for CSR

'The board are highly supportive and do not ask how much time and money is invested in it. Every single person that's ever been involved in the board gets caught up in the enthusiasm as well.' Having said this, there are emerging concerns within the company about the amount of time spent by the MD in promoting it. Nevertheless, there has been no systematic attempt to measure this.

External barriers to adopting CSR

'There is a "tall poppy" syndrome. Stick your head up above the parapet and do something different and there is a risk of being shot down. You are also on your own, since governments show no leadership and are driven more by public opinion than by supporting good ideas ... they don't want to pick winners but are more than comfortable to accept the trophy on behalf of your hard work'. Another barrier is persuading other business partners to engage with the message, idea and vision – 'CSR is only as strong as the number of companies that are involved. It is something that is done with business partners. I focus on engaging at the CEO level because they can just make a decision. It can be difficult to get the idea across since you are talking about qualitative outcomes. But once people understand what its about they are locked in. The connection with children is incredibly strong. It's emotive and everyone can relate to it.'

Internal barriers to CSR

One of the main internal barriers is justifying ROI. 'You have to justify that any money you spend on this is giving you a return. This is a barrier because the ROI isn't clear. We see REA as a marketing initiative. It fits into our marketing budget and strategy and we just accept that 50 per cent of what you invest in marketing is probably lost and we accept this because you never know which 50 per cent is working.'

While the board is supportive of CSR, it is clearly seen as a high-risk activity since there is no way to measure precisely the ROI. So this means that people have to believe in it for many reasons, not just financial – 'It's also essential to get staff involved otherwise it's just an add-on and won't fully benefit the business and will die'.

Developing a CSR culture

CONCENTRIC's culture from the start has been based on a tribal model – 'It means that the company has no hierarchy, no titles and people go out and hunt in teams and bring back resources to the family'. It is claimed that this mental model allowed the CSR culture to take hold very easily – 'REA is run by people in the office. It's run by staff and owned by them. All the staff have become involved. They love to meet the kids. It's something that defines the company and is a core part of the experience of working there'.

The main challenges in successfully implementing CSR strategy

The main challenge is time. REA has taken a life of its own and there is a need to keep it under control while at the same time maintaining commitment – 'Without this it would dwindle and die. You must keep feeding it. You need to market the benefits all the time'. It also comes down to leadership and having a breadth of goals which allows a company to pursue a CSR agenda. The MD's opinion was that companies that are 'narrowed in' on a few outcomes such as profit find it very difficult to accept a CSR agenda since it doesn't fit into the rigid vision of what success means and how it is defined. Also, they tend to be less innovative and more risk averse in experimenting with rather nebulous initiatives such as CSR where there is a leap-of-faith involved – 'We are not like this. Our goals have always been very broad. Much broader than just making money. We now apply in the office the same learning strategies we use with the students and the result is seen directly in increased self-efficacy and innovation which we can measure'.

Measuring ROI

ROI is not measured in any hard, formal sense in CONCENTRIC – 'There is an awful amount of anecdotal stuff that comes from people in the company that just indicates it works. If we put it on a spreadsheet then we would stop doing it. We discuss it at board meetings but have never attempted to add up the hours we put into it – 'If I did this my accountant and my wife would shoot me'. There was a strong view that a spreadsheet cannot capture the soft benefits which is where a lot of the value lies and there is a need to just accept that like any marketing initiative ROI is difficult to measure – 'We monitor it intuitively and anecdotally every week in our senior management meetings because we know that many of the benefits just can't be measured. For example, when I get asked to stand up and talk about the REA at major industry conferences, I ask myself "what would I have had to pay for that opportunity?". I think we would be scared if we actually added up how much time I invest in REA. Yet we have no empirical evidence that it impacts on sales and we have never tried to collect it. We do try and assess what we get out of it but it's all very subjective'.

The main perceived benefit is CSR's impact on customers' perception of CONCENTRIC's business and what it stands for – 'It's given us a profile we could have never had. One which our competitors are scared of because we get such high profile coverage and exposure and they simply can't compete ... We ran the national finals of the REA at Parliament House and it was Christmas and all the politicians walked in and were stunned by what they saw. There is also a magazine that goes around inside Parliament House called *The House* which documents the subcommittees and all the work that goes on and we had a whole article dedicated to the REA. We are trying to get an 'Australian Story' or 'Four Corners' done of REA ... If we can make this happen most of the messages can be very subliminal'.

There is also a perceived increase in employee satisfaction and the ability to open doors into the boardrooms of some of the world's biggest companies that would have otherwise been closed. And there is a belief that business partners also benefit since they are drawn into the process – 'everyone wins and it strengthens the partnerships and loyalty in my supply chain which probably gives us other benefits in the long-term'. Finally, REA has also created a perception in CONCENTRIC's labour market that they are a firm that people want to work with – 'The things that they say in interviews is that they like the culture best. We know the lecturers at various universities and they tell the best students about us. We have them lining up at the door. We don't have to advertise. We are perceived to be a better organisation and through this we tend to pick up really good staff and like-minded clients who want to achieve the same goals'.

Reporting CSR successes and failures

As CONCENTRIC is not a listed company there is no formal reporting requirement. CONCENTRIC does, however, formally report the activities of the REA to the board every month. The REA also produces newsletters and media stories to update its stakeholders – 'We try to keep them all involved since they are a part of the initiative and key to its success, so external and internal reporting is just as important'.

Key determinants of CSR success

CSR needs to be made part of our core strategy and corporate goals at a high level. It is also important to avoid being too narrow with corporate goals otherwise CSR will not fit comfortably within the overall corporate strategy. Staff need to be involved and it takes 'pig-headed determination' to make it work, which in turn requires a long-term commitment to the cause being pursued. It's got to be driven by personal belief, determination and continuous effort. It is also about linking with the right people who can see the bigger picture and linking to the idea of heroes. CSR is also a high risk activity, so it is important to ensure that risk is managed effectively – 'You have to be careful you don't have a failure because while the benefits are high the potential downside is also quite high especially because we are dealing with kids'.

Crown Project Services

Crown Project Services (CPS) is an independent project-management consultancy with over 40 staff which provides services in project management, programme management, auditing and event management. CPS service a wide variety of industry sectors including: commercial and industrial; tourism, hospitality, sporting and leisure; government, education, health and justice; and infrastructure, ports, rail and green technologies.

Drivers of CSR

CSR in CPS has been driven from the top-down – 'It is the coalescence of the aspirations of the directors of the business. We are the drivers of the strategy'. The inspiration for CPS's CSR activities arose from their involvement in the Sydney 2000 Olympics construction programme where they saw the benefits of environmentally sustainable development (ESD) and of setting the moral high ground. CPS appears to have carried those thought processes over to their CSR strategy – 'It's driven by public service principles and we wanted to push the boundaries of ESD and also behave ethically in everything we do ... We only tender on what we see as good social outcome projects'. Personal experience of the Asian Tsunami by the two founding partners also led directly to the formation of the SkyJuice™ Foundation. This not-for-profit charitable organisation is another strand of CPS's CSR activities which provides sustainable water solutions for humanitarian and disaster relief (www.skyjuice.com.au).

CSR strategy

CPS's CSR strategy is largely based around the SkyJuice™ Foundation and its sustainable development activities. It is a differentiation-based strategy rather than price-based strategy and has forced CPS to be selective about business partners and the services it provides – 'CSR focuses its business on certain types of clients that think long-term, have similar values and don't force us to compete on price alone'. However, the CSR strategy is also by necessity strongly pragmatic because of the challanges in persuading some CPS clients to invest in ESD – 'Our strategy is not to overtly promote CSR as part of our business but it has become an integral part of who we are and what we do – particularly our attachment to the principles of ESD. The Foundation is kept quite independent of the business'. Although the Foundation is not used as an overt marketing strategy it is often discussed informally with CPS clients and there is a view that it has 'brought kudos to the company' and generated some business opportunities overseas'.

Who was involved in the development of the strategy?

CPS's CSR strategy has always been driven by the four main shareholders in the business who are the two founding partners and their wives – 'We were all inextricably linked with this Tsunami experience which brought about a forced change of mind'. The SkyJuice™ Foundation is governed by eight founding members (including the four company directors) who are responsible for ensuring that the Foundation adheres to its articles of association. CPS staff were initially involved in the SkyJuice™ foundation although now it is more distant. It is certainly not a formal aspect of their role in the business – 'they get a warm feeling from it'. However, staff *are* intimately involved in the ESD part of the CSR strategy through their voluntary involvement with the Green Building Council, strong expectations from directors that they promote ESD and feedback from clients which provides valuable feedback for incorporation into future strategy.

Board support for CSR

'We all personally believe in ESD and we are all personally involved in the SkyJuice™ Foundation. We all contribute funds and time and energy to support what we do in this area. Ultimately, our job is to persuade our clients that ESD is a worthy investment and we do that with personal passion'.

Barriers to adopting CSR

CPS identified a series of internal and external barriers to CSR – 'Internally there will always be the sceptics but there should be none if CSR is driven from the top-down and if one employs the right people who also believe in its merits'. Apart from this there are no major internal barriers. Externally – 'Our clients are the only barrier and we have to educate them through personal interface and repetition. We have taken one client from an absolute disbeliever to interest. We can't educate the industry'. The suggested reason for client scepticism is that CSR is perceived to be an inherently subjective and risky activity which requires a multidisciplinary approach which is 'too often lacking in the industry'. There is also frustration that decision-makers in client organisations are more often than not, unprepared to take the risk of investing in CSR type initiatives (such as environmentally and socially sustainable designs) because ROI of CSR is not easily measured and there may be a long-year-payback period. This means that selling CSR involves personal commitment on the part of employees and trust on the part of potential clients – 'the board's don't care and are not interested because they are only interested in the metrics and in three-year-payback periods ... CSR is long-term and boards tend to focus on the short-term, so persuading them to invest is difficult sometimes. If boards were more prepared to back decision-makers who are willing to take a risk with innovation and something new then it would be much easier. It must be driven by board culture and from top-down, it can't happen bottom-up ... There is a lot of trust in selling CSR and our staff have to know what they are talking about and sell it with commitment'.

The term CSR itself is also seen to be problematic in that it is too emotive and intangible – 'If we talked about CSR, clients would be sceptical, but if we talk about the dollar benefits and earning potential to their business and talk about CSR as a nice add-on then it works. But they will drop it like a hot potato if their profit starts to fall. So we have tried to put some statistics behind the innovations we offer but they are ten-year-plus-payback periods and frequently they don't get a guernsey because to get the board's attention they have to be five years or less'.

Other external barriers to the adoption of CSR include lack of regulatory incentives and guidance, long-payback periods, a lack of foresight in the industry and a lack of public pressure for firms to engage in such activities – 'Only when ESD is mandated by government does anything happen and there has to be public sector leadership. Without this we would still be stuck back in the dark ages. It's got to be driven by government policy and government leadership. We find it very hard to find government policy to justify what we want to do. So we have to drive ESD from a moral foundation'.

Developing a CSR culture

Persistence and embedding CSR are identified as key to developing an appropriate culture – 'You have to keep saying it and people will eventually get the message. ESD is an integral part of everything we do and we talk about SkyJuice™ in our whole-of-company get-togethers'. To build a CSR culture, CPS has also incorporated ESD into performance reviews and employees are trained through the Green Building Council – 'We know their thoughts and they know ours because the executive team do all the performance reviews which encourage staff to engage in innovation and extra-curricula activities which are relevant to our firm'. While there is a preference to recruit people with common beliefs, the approach to CSR in CPS is a very pragmatic one because in a small firm CSR is considered a luxury compared to other core competencies – 'If they are not interested then fine, we can't thrust it down their throat ... It's difficult to get good people so it's difficult to build CSR into our recruitment and if a candidate rabbitted on about CSR then we would be sceptical since we would think we didn't have a doer. At the end of the day you have got to earn a quid. You have to deliver first and if you, during the delivery, add extra value to the long-term assets of the community then that's great'.

Challenges in successfully implementing CSR strategy

The implementation of the CSR strategy is organic and 'implements itself'. The challenges are also mainly external and relate to government policy and client attitudes towards ESD – 'To engage with CSR we have to be able to persuade our clients that it will benefit their business. We have to be able to sell the corporate and personal benefits of innovation. This is not always easy because there are no external incentives for them to invest in ESD. The challenge is education of the client base'.

Measuring ROI

ROI associated with CSR cannot be measured by normal financial metrics and there is no systematic attempt to do so – 'You will hardly ever get financial return on the human investment involved. You wouldn't go down this route for a straight, boring financial return. There has to be a passion for it'. There is also an internal and external dimension to ROI of CSR which appears to be intimately linked. The external dimension is associated with the ROI for CPS clients and there is also the internal ROI for the business. Again reference is made to the issue of trust which implies that it is easier to sell the benefits of CSR to long-term clients and customers – 'At the end of the day we are a service-based company managing cash on behalf of our clients. We try to spend their money as wisely as if it were their own with a CSR hat on. There is no cost to us of CSR and some clients just flatly refuse our innovations for the reasons discussed. For our clients, there is a ten-year-payback period. The benefits we sell to clients of engaging with ESD are lower emissions, increased staff safety, satisfaction, retention and productivity, reduced energy bills, reduced recurrent maintenance costs, longer asset life and better serviceability. But too often clients go for the short-term dollars and to persuade a client of the benefits there is a lot of trust needed in us. There has got to be trust'.

Reporting CSR

CPS is not a listed company and have no formal reporting requirements but it does produce an annual report which is not published and where CSR initiatives are reported, mainly subjectively – '... triple bottom line reporting has not got as much credibility as the theoreticians claim it has. There is a lot of subjectivity'.

Key determinants of CSR success

The main determinant of CSR success is clarity and objectivity about the benefits to employees, communities and clients about CSR – 'Be as objective as you can. Try and use CSR to get a competitive edge over your competitor by being a better corporate citizen than them because the only other thing you have is lower price'. CPS refrains from engaging with clients where it is forced to compete on price alone. Finally, while objectivity is important – 'without the passion it's easy to drop CSR'. It would seem that although CSR needs to be driven by a personal belief and passion, this must be rationalised when dealing with clients to avoid being seen as evangelical.

Conclusion

Within the limited range of CSR case studies and approaches that are presented in this chapter, there seems to be a clear indication that most of the firms we interviewed associate CSR with ESD and the need to do or to be seen to do, in their own words, 'the right thing' – a deontological (duty-based) ethical philosophy. The similarity in language used and the significant overlap in approaches in each category also suggests that the field of CSR is in its early stages of development in the construction and engineering industries and that firms are searching for a way to differentiate themselves in this space and to thereby derive competitive advantage. The distinct dichotomy between what public listed companies and privately owned companies do when it comes to CSR, is also not entirely surprising and is borne out by the literature which suggests large companies generally have developed a more systematic and strategic approach to CSR compared to private companies and SMEs which adopt a more informal, ad hoc and un-strategic approach. SMEs do not yet seem to see the necessity of adopting a formal CSR strategy – the kind that at least requires a 'CSR conducive' governance framework, specific reporting regime and a deliberate shift towards a CSR culture – as a prerequisite to gaining competitive advantage within the industry. While CSR, to a certain extent, does appear to play a role in product and service differentiation between firms, the core day-to-day operations of many SMEs seem to remain significantly unaffected by the CSR movement. In fact, they claim that they have been practicing CSR without needing to label, formalise or market it as such.

6 Conclusion

In this chapter we revisit the original aims of this book and draw together the lessons from each case study. Combined with the arguments we have made throughout this book, we conclude by proposing an alternative approach to CSR strategy which can potentially provide strategic advantage and challenge the compliance-based methods which currently dominate the industry.

Introduction

We started this book with the simple premise of providing a balanced and critical debate on CSR and discussing its strategic role in the construction and engineering industries. With so much being written about the integral role of business in advancing the well-being of societies and the environment, we felt it was difficult for readers to separate out the rhetoric from the reality in the CSR debate. We also felt that while social, political, institutional and market-driven pressures for greater CSR are real and likely to intensify in the foreseeable future, there seemed to be a disjoint with where the CSR debate was heading. Presently, the growing literature advocating and philosophising the need for firms to adopt some form of CSR initiative is not matched by evidence-based insights into how firms should strategically pursue and operationalise CSR as a means of achieving sustained competitive advantage. This is unfortunate and there are three main reasons why this does not help the CSR cause.

First, to all intents and purposes, firms will be more inclined to incorporate CSR strategically into their business if there is proof that it can be used to improve their financial bottom line, or more broadly improve their competitive edge (whichever way that is defined or measured). If, by doing so, firms can achieve the triple bottom line of advancing economic, societal and environmental good then so much the better, but to sell the CSR argument purely on the grounds of doing good without any strategic purpose is naïve and impractical for the reasons we have discussed throughout this book. Our case studies reinforce this view in the sense that firms do indeed see CSR as having strategic benefits, even if not always formalised, which can bring about certain tangible and desirable advantages to their businesses.

The second reason why the growing CSR literature does not help advance the CSR cause relates to the multitudes of exaggerated CSR success stories, often aptly referred to as 'green wash', which create a déjà vu of the 1980s and 1990s experience of new management fads. In most instances, these were more effective at improving the profitability of management consultancy firms than the performance of business and the well-being of those who worked for them. We have shown in this book that it is naïve to pitch CSR as a wholesale solution for all firms to do well. Indeed, it would damage the CSR cause to do so. For many firms, it does not make business sense to simply jump on the CSR bandwagon without truly understanding the specific business environment in which it will be applied, and the demands and implications it may have for resources and, consequently, the long-term health of the business. We have stressed that a highly formalised CSR strategy does not necessarily suit every firm and not every firm needs to go down this route in order to be successful. To suggest otherwise would be misleading and counterproductive because an overly prescriptive approach ignores the range of possible prevailing firm-, industry- and country-specific factors and circumstances that might impact on any firm at any particular time and, hence, determine the most suitable type of CSR to adopt and how best to operationalise it. The case studies have reflected clearly that each firm has their own interpretation of how and why CSR should be practised and at what pace it should be implemented. We believe that recognising this, rather than dictating standardised models and so-called 'recipes' for success, is the most appropriate and practical way forward. Rather than viewing this as a shortcoming due to a lack of a standardised CSR approach, we argue that businesses should be encouraged to be creative and innovative about how best to fit CSR into their business without getting too preoccupied with the notion that there is an ideal CSR strategy.

The third reason why the growing CSR literature does not help advance the CSR cause is the failure to connect it convincingly with improved business performance. For the majority of firms in the construction and engineering sector, the connection between CSR and business performance is uncertain and problematic, ensuring it continues to be seen as something needed for regulatory/legislative compliance, rather than strategic value. We have discussed how this challenge might be addressed by way of: creating an appropriate vision, mission and governance structure; developing realistic, clear and measurable goals in partnership with key stakeholders; creating business indicators and reporting mechanisms that capture the critical performance measures; and building a conducive organisational culture that supports CSR. However, while these may be key elements of a successful CSR strategy, our case studies have also shown how different firms have responded to the challenge in very different ways and the lessons they have learnt in doing so. In the remainder of this chapter we will distil these lessons into three firm categories – *large developers, large consultants* and *smaller firms* in the supply chain. In doing so, we acknowledge the limited size of our sample and the need for more in-depth academic research. However, we do believe that the lessons learnt provide useful points of reflection for managers contemplating the merits of CSR in their unique firm and business environment.

The drivers of CSR strategy

For the large developers the main drivers of CSR were: new legislation; investors; competitors; employee demands; clients (particularly government); tenants; increasingly vocal communities; and the emergence of new industry tools at a global and domestic level which enable them to measure and monitor progress against accepted goals. While CSR has been a compliance–driven business imperative in the past, it is increasingly seen as a potential mechanism to achieve competitive advantage, particularly through increased trust and reputation with stakeholders. ESD and 'Doing the right thing' were common slogans used in these large developers and there was a considerable degree of uniformity in approach, in what can be described as a fairly small and tightly knit 'sustainability' community. There was not much to differentiate the strategies of those developers we interviewed, despite the term 'competitive advantage' being widely used. The real argument being proposed in these developers is that by balancing economic, ecological and social imperatives, organisations can not only experience direct short-term operational savings in energy consumption, etc., but importantly, they can also 'future proof' their business by better managing risk, by being more innovative and building more long-term value, resilience and longevity. So at the moment, competitive advantage from CSR would appear to be price-based rather than differentiation-based, and driven by potential operational savings rather than from the differentiation of services and products.

For the large consultants the main drivers of strategy were their values and the personal commitment and vision of their owner(s). External regulatory pressures featured as a less influential driver than in the large developers although the clients of large consultants are also increasingly requesting greater attention to environmental issues in particular. Social issues remain largely irrelevant to clients because they are not legislated in the same way as environmental issues. For large consultants, CSR is not a necessity (as in the case of large developers) but an increasingly important aspect of their business, which is needed in response to a resource-constrained and regulated world of more informed and empowered stakeholders. CSR is also seen as a potential source of competitive advantage and a way to exploit new business opportunities presented by challenges such as climate change and carbon trading. It is also seen as a way of creating a formal framework for capturing, structuring and communicating what they had been doing in an ad hoc manner for many years.

For the smaller firms further down the supply chain, CSR was not seen as something that needed formalising but as something which was inherent in a 'family' or 'tribal' business model which relies on perceived long-standing and intimate relationships with local communities and employees. It is argued that the drivers of CSR have always been present in these small firms which appear to be relatively unconcerned with the regulatory imperatives that are driving the larger firms (at least for the time being). Not surprisingly, the firm's owner has 'ownership' and close control over strategy, which is likely to be driven by their personal experiences and preferences for different causes and historical connections with communities. The dollar benefits from CSR were particularly important and there was a strongly assumed, although not formally acknowledged, link between CSR and business success.

The nature of CSR strategy

For the large developers, CSR strategy is formally set out in an annual report with clearly defined objectives and measurable KPIs guided by globally established frameworks such as GRI. Typically, the highly structured strategies are a work in progress and part of a rolling longer-term plan (usually around five years) to balance economic, ecological and social objectives. Although to date, there has been a strong focus on ecological and, more recently, social imperatives, CSR is mainly built around the dollar benefits that could be derived from it. CSR in these large organisations depends on consistent procedures and systems which enable integration across business units (often with very different organisational and national cultures), building alliances through the supply chain and encouraging employees to uphold a clearly articulated set of core business values.

The large consultants are typically in an earlier stage of development in formalising CSR than the large developers although they also recognise the potential business benefits from doing so. Guided by more limited resources and a project-based culture, CSR is typically built around a relatively limited set of voluntary initiatives to help achieve a simple company vision and a set of corporate goals. The key to CSR for these firms is achieving focus, balance and scale – addressing emerging needs without compromising the future of the firm.

The strategy of smaller firms is typically far more reactive than proactive. It is often not driven by any clear objectives but by a response to ongoing relationships with causes or new opportunities which present themselves. Strategy is likely to revolve around very few or even a single initiative and tends to develop in an organic and almost spontaneous fashion rather than being pre-planned and strategic. CSR is largely undocumented and is something the business takes for granted, being highly pragmatic and not-marketing driven (at least in an overt sense). Initiatives can often take a life of their own, resulting in spin-off businesses and are often seen as a way to build closer relationships with clients and business partners.

Who is involved in developing CSR strategy?

For the large developers CSR is typically, although not always, driven by a multi-layered and hierarchical committee structure which ultimately is responsible to the board. Strategy is normally driven by the board and a senior sustainability manager whose job it is to integrate sustainability into the business, report to the board and lead a sustainability group or committee with representatives from different parts of the business. This high level committee is in turn typically advised by steering and working groups led by specialist sustainability managers throughout the company whose responsibility it is to drive sustainability and liaise with external bodies who can help to inform strategy and provide practical tools to measure and benchmark performance.

The large consultants tend to avoid complex hierarchical structures but instead establish a board subcommittee or separate unit supported by sustainability working groups in each business which are led by sustainability coordinators.

For smaller firms the development of a CSR strategy is normally led by the personal motives and interests of the managing director. Employees are typically involved but not through any formal committee structure.

Barriers to CSR strategy

For the large developers the main barrier to developing a CSR strategy is through established cultures and ways of working. There is also a need to address potential misunderstanding and confusion of what CSR means for a large and diverse business. Since it requires people to think differently, resourcing and education are also critical, as is finding leaders and champions with the vision, leadership and management skills to bring about cultural change. Another barrier is scepticism in some quarters about tangible monetary benefits to the business and its bottom line. CSR is seen as risky because its benefits are not supported by hard data and ROI is difficult to measure. So overcoming the fear of the unknown and reducing the apparent risk associated with CSR is important. This requires advocates to be factual and rational rather than emotive and evangelical, to be as pragmatic as possible and to ensure that any arguments are evidence based. Getting reliable data to support strategy is therefore a critical challenge which is not helped by the lack of regulatory guidance in this area. Implementing strategy throughout long and unwieldy supply chains of smaller firms which may be resistant to the idea is also seen as a major barrier to implementation.

For the large consultants which operate in a highly project-based environment, CSR often depends on the goodwill of individual employees working outside the boundaries of projects. This can make it difficult to maintain momentum. Some clients can also be potential barriers since not all are willing to invest in something which lies outside their normal investment horizon and which may only produce a long-term return or no apparent return at all. Hence, being able to persuade and educate individual clients about the merits of CSR is important to effective implementation.

For the smaller firms, CSR is often deemed as a highly risky process. CSR is usually seen as an unaffordable long-term luxury in a cash-flow driven business. For this reason, ROI must be obvious and relatively quick for any initiative to be adopted and the same also goes for clients who may be reluctant to invest, particularly with the lack of regulatory incentives from government to do so. Another potential barrier is that CSR often requires a multidisciplinary approach which is often missing in a small business.

Building a CSR culture

For the large developers, building a CSR culture is a gradual and challenging process because of the diversity and size of their organisations. This depends upon senior management not only leading by example but promoting the importance of responsible behaviour by incentivising staff. Having a defined strategy underpinned by clear values and measurable objectives is important, as is communicating that strategy clearly to those who have to implement it. Since CSR requires a change in behaviour, the ongoing education of staff about the personal and organisational benefits of responsible behaviour is critical. This is commonly achieved through ambassadors and demonstration projects (at least at first) and in the longer-term, building responsible-behaviour objectives into individual job descriptions and performance-evaluation processes and rewarding individuals on the basis of achieving them. In the long-term, dispersing responsibility for good behaviour in this way is a far more effective way of building a CSR culture than directing it under the auspices of a separate sustainability unit or department.

For the large consultancies and smaller firms, culture change tends to occur more by osmosis and mentoring than by a structured process of performance management and monetary incentives. The greater intimacy of these firms, often engendered by their governance structure and size, makes it particularly important for staff to understand clearly what responsible behaviour means to their firm, how it specifically applies in their business and how it relates to the firm's core values, which tend to be more internalised than in the larger firms. To assist in this process, staff themselves may be intimately involved in establishing and managing the relevant initiatives. In this way, while strategy, leadership and senior management support is important, culture change is more likely to be a democratic process than a top-down driven process as in larger firms.

The challenges and determinants of success in developing a CSR strategy

For the large developers the main challenge is implementing and managing cultural change. This depends on a wide range of factors but ultimately on genuine and widespread senior-management support, leadership and commitment which is underpinned by a willingness to take risks and to innovate. Senior managers must talk about CSR at every opportunity and lead by their own example. Also important is redefining people's roles and responsibilities and linking individual rewards to clearly defined and measurable goals. This process of cultural change must be paced with sensitivity to the maturity and culture of the organisation, since taking an unprepared organisation too far too fast will likely create cynicism and resistance to change. All of the above requires continuous internal and external communication of the benefits of responsible business practices. This is best achieved when it is possible to demonstrate ROI and, in particular, a positive relationship between financial performance and CSR. However, the direct relationship between performance and CSR is not easy to establish. Furthermore, many of the advantages of CSR are qualitative and firm-specific and therefore difficult to measure (reputation, ethical practice, stakeholder engagement, etc.). Despite these challenges, there is a growing need to clearly demonstrate ROI which in turn requires clear measurable targets. To this end, large developers tend to subscribe to national and global indices, KPIs and tools to legitimise their strategy and to report them formally in increasingly detailed annual or sustainability or corporate-responsibility reports. However, there has been far more progress in measuring environmental performance than social performance, largely because of the compliance imperative in this area. Another major challenge in large firms is bringing about culture change across large and diverse organisations and supply chains which are often geographically and functionally dispersed. It is crucial to establish a common understanding and vision across a large number of people of what CSR means for the organisation and why it is important at an organisational and personal level. Empowerment, involvement, reward systems, education and effective communication (informal and formal) is key to this, because unless CSR is real and meaningful for individuals, initiatives will not be adopted. All of the above requires continuous internal and external marketing of the benefits of responsible business practices while at the same time avoiding any 'green wash'. A final and significant challenge for large developers is embedding and integrating CSR into the business and establishing systematic and consistent systems and procedures across different business units and regions. This not only creates a common language to discuss CSR but enables ROI to be measured more effectively and consistently across the business.

For large consultants success ultimately depends on the personal vision and determination of the MD or CEO. It is also important that the CSR aligns with broader organisational values. Developments in formally defining, measuring and reporting ROI are less advanced than in larger firms. In contrast to the large developers, a highly systematic and formalised approach is seen as a disadvantage in its potential to undermine the sense of spontaneity which characterises CSR in smaller firms. Another challenge is maintaining momentum in a project-based business. In this environment, initiatives depend on the goodwill and personal passion of staff that drive them, often outside normal working hours.

For the smaller firms the main challenge is maintaining pragmatism and managing scope-creep. CSR is potentially a 'bottomless pit' which can create an unsustainable drain on limited resources. Relative to larger firms, the human and financial investment can be enormous and there is a real danger that initiatives take on a life of their own and develop drivers beyond the core objectives of the founding firm. Another key challenge is educating clients about the merits of investing in what may be longer-term initiatives that may not produce obvious and immediate returns. ROI is rarely formally measured (at least in precise terms) but instead broadly and subjectively judged from anecdotal stories about the benefits for the business.

Conclusion

The approach to CSR we have advocated throughout this book is different from the compliance-based approaches to CSR which dominate the construction and engineering industries. A compliance-based CSR strategy is based on a shareholder-primacy model of governance and is mainly driven by narrow and short-term financial business objectives. At a human level, it is also driven by a belief that people will always act in their own interests and are best prevented from acting unethically by top-down imposed prescriptive codes of conduct, close auditing, monitoring and control, and by threats of economic sanctions. In contrast, our case studies and the approach we advocate for CSR strategy recognises that firms serve people who have a more complex set of needs and motivations, other than economic self-interest, which must be understood and considered. Our approach advocates both emergent and pre-planned components in strategy and relies on self-imposed values becoming the guiding principles for behaviour rather than some externally generated guidelines and standards. In this regard, we refer to a growing body of literature and evidence that show how companies can enact a strong organisational identity and business culture as a tool to create, harness and shape the norms, values and belief systems which underpin successful CSR. We believe that this is the best way for firms to derive genuine differentiated competitive advantage from CSR, rather than just short-term cost advantage. We also believe it is the best way for firms to ensure organisational responsiveness and resilience to the many new ethical dilemmas that will inevitably face construction and engineering firms into the future. This will often fall outside the predetermined scenarios and responses that represent the basis of newly enacted laws and regulations. Finally, we also believe our approach will avoid the potential pitfall, that in securing external verification of compliance, managers will misconstrue this to mean that they are genuinely challenging their underlying values and convictions.

While we have argued throughout this book that a firm's approach to CSR is contingent upon its unique environmental context, size and maturity, etc., and that there is as yet no definitive link between this and better performance, we do assert that if firms do not move beyond a compliance-based approach their long-term success and profitability will likely suffer and the potential competitive advantage to be gained will probably never be realised. Although it seems that firms are not richly rewarded in monetary terms for investing in CSR at the moment, we argue that such investments will help bring about long-term shareholder value. As pointed out by researchers, the strength of CSR may not lie entirely with its economic advantages but with its effectiveness through enhanced legitimacy of corporate activities. Hence, we would argue that better or cleverer approaches to CSR, that are integrative with the core business strategy, may be the key to producing better returns in the future.

Having said the above, we have also recognised that this approach is difficult to implement in practice, particularly within the construction and engineering sector. The picture that emerges from this book is of an industry that is primarily motivated by profit and which is best moved by regulation. Although it has begun on a path of engaging in CSR, it is still a largely compliance-driven industry that is unlikely to move beyond the minimum requirements without external pressure and a clear business case. It is also a risk-averse industry which serves risk-averse clients and which has little trust in the many subjective arguments that are made to support the case for CSR. Until this culture changes (primarily through education of current and future employees and clients), the idea of behaving responsibly will continue to mean achieving economic success, with ecological goals a second regulated priority and social goals a distant third.

Bibliography

Abramson, P. R. and Inglehart, R. (1992) 'Generational replacement and value change in eight western European societies', *British Journal of Political Science*, 22: 183–228.

ACCA (2003) *Corporate social responsibility: is there a business case?*, London: The Association of Chartered Certified Accountants.

Aguilera, R. V. and Jackson, G. (2003) 'The cross-national diversity of corporate governance: dimensions and determinants', *Academy of Management Review*, 28: 447–465.

Ansoff, H. I. (1965) *Corporate strategy*, New York: McGraw-Hill.

Arnstein, S. R. (1969) 'A ladder of citizen participation', *Journal of the American Institute of Planners*, 35 (4), July 1969, 216–224.

A.S.X. Corporate Governance Council (2003) *Principles of good governance and best practice recommendations*, 1st edition, Sydney: A.S.X.

Babeau, O. (2007) 'Granting disorder a place in ethics: organisational deviance practices and ethics', in Carter, C. S. C., Kornberger, M. S. L. and Messner, M. (eds) *Business ethics as practice,* Cheltenham: Edward Elgar Press.

Bailey, W. and Spicer, A. (2007) 'When does national identity matter? Convergence and divergence in international business ethics', *Academy of Management Journal*, 50, 1462–1480.

Banerjee, S. B. (2007) *Corporate social responsibility – the good, the bad and the ugly*, Cheltenham: Edward Elgar Press.

Barnett, M. L. (2008) 'Stakeholder influence capacity and the variability of financial returns to corporate social responsibility', *Academy of Management Review*, 32, 794–816.

Barnett, W. P. and Burgelman, R. A. (1996) 'Evolutionary perspectives on strategy', *Strategic Management Journal*, 17, 5–19.

Barnes, P. (2002) 'Approaches to community safety', *Australian Journal of Emergency Management*, Autumn 2002, 15–23.

Baron, F. (2007) 'Lessons from the UK', *Infrastructure magazine*, 50–54.

Bartholomeusz, S. (2007) 'Institutions muscle up over governance', *Sydney Morning Herald*, 20 April 2007, 20.

Basu, K. and Palazzo, G. (2008) 'Corporate social responsibility: a process model of sensemaking', *Academy of Management Review*, 33, 122–136.

BCA (2007) The centre for corporate public affairs, Business Council of Australia, Sydney, Australia [online]. Available at: www.bca.com.au

Beatty, A. (2009) 'Carbon reduction: far more than compliance', *Risk Magazine*, March 2009, 15–17.
Beder, S. (1998) *The new engineer: management and professional responsibility in a changing world*, Melbourne: Macmillan Education.
Bettis, R. (1998) 'Commentary on "redefining industry structure for the information age" by J. L. Sampler', *Strategic Management Journal*, 19, 357–361.
Bies, R. J., Bartunek, J. M., Fort, T. L. and Zald, M. N. (2008) 'Corporations as social change agents: individual, interpersonal, institutional, and environmental dynamics', *Academy of Management Review*, 32, 788–793.
Black, R. (2007) 'PPPs: the global scorecard and the Australian Experience', *Infrastructure Magazine*, 7, 19–20.
Blair, M. M. (1995) *Ownership and control: rethinking corporate governance for the twenty-first century*, Washington, DC: The Brookings Institute.
Bourguignon, A (2007) 'Are management systems ethical? The reification perspective', in Djelic, M. and Vranceanu, R. (eds) *Moral Foundations of Management*, Cheltenham: Edward Elgar Press, 212–243.
Boyce, G. (2000) 'Public discourse and decision making', *Accounting, Auditing and accountability*, 13 (1), 27–64.
Braithwaite, J. and Drahos, P. (2000) *Global business regulation*, Cambridge: Cambridge University Press.
Brown, J. and Fraser, M. (2006) 'Approaches and perspectives in social and environmental accounting: an overview of the conceptual landscape', *Business Strategy and The Environment*, 15, 103–117.
Brown, M. (2005) *Corporate integrity: rethinking organisational ethics and leadership*, Cambridge: Cambridge University Press.
Burby, R. J. (2001) 'Involving citizens in hazard mitigation planning: making the right choices', *Australian Journal of Emergency Management*, Spring 2001, 45–48.
Camenisch, P. F. (1987) 'Profit: some moral reflections', *Journal of Business Ethics*, 6, 225–231.
Cameron, K. and Quinn, R. E. (2005) *A framework for organisational quality culture*, New York: ASQ.
Campbell, J. L. (2008) 'Why would corporations behave in socially responsible ways? An institutional theory of corporate social responsibility', *Academy of Management Review*, 32, 946–967.
Carroll, A. B. (1991) 'The pyramid of corporate social responsibility: towards the moral management of organisational stakeholders', *Business Horizons*, July/ August 1991, 39–48.
Carter, C. and Lorsch, J. (2004) *Back to the drawing board: designing corporate boards for a complex world*, Boston, MA: Harvard Business Scholl Press.
Carter, C., Clergg, S., Kornbereger, M., Laske, S. and Messner, M. (2005) 'Introduction', in Carter, C., Clergg, S., Kornbereger, M., Laske, S. and Messner, M. (eds) *Business ethics in practice: representation, reflexivity and performance,* Cheltenham: Edward Elgar Press.
Chandler, A. (1962) *Strategy and structure*, Cambridge, MA: MIT Press.
Chang, C. Y., Chou, H. Y. and Wang, M. T. (2006) 'Characterizing the corporate governance of UK listed construction companies', *Construction Management and Economics*, 24 (6), 647–656.
Charan, R. (2005) *Boards that deliver*, San Francisco: Jossey-Bass.
Charkham, J. (1995) *Keeping good company*, Oxford: Oxford University Press.

Cheah, C. Y. and Gavin, M. J. (2004) 'An open framework for corporate strategy in construction', *Engineering, Construction and Architectural management*, 11 (3), 176–188.

CIOB (2006) *Corruption in the UK construction industry: survey 2006*, Ascot: The Chartered Institute of Building, p.3.

Clarkson, M. B. E. (1995) 'A stakeholder framework for analyzing and evaluating corporate social performance', *Academy of Management Review*, 20 (1), 92–117.

Cole, T. R. H. (2003) *Final report of the Royal Commission into the building and construction industry: summary of findings and recommendations*, Canberra: Australian Government.

Cummings, S. (2005) *Recreating strategy*, London: Sage.

Dainty, A., Green, S. and Bagilhole, B. (2007) *People and culture in construction – a reader*, London: Taylor & Francis.

Deakin, S. (2005) 'The coming transformation of shareholder value', *Corporate governance: an international review*, 13 (1), 53–67.

DIAC (Department of Immigration and Citizenship) (2009) *Population flows: immigration aspects – 2007–2008*, February, Canberra: DIAC.

Dimaggio, P. J. and Powell, W. (1983) 'The iron cage revisited: institutional isomorphism and collective rationality in organisational fields', *American Sociological Review*, 48, 147–160.

Dimento, J. F. C. and Gies, G. (2005) in Tully, S. (ed.) *Research handbook on corporate legal responsibility,* Cheltenham: Edward Elgar Press.

Donaldson, T. and Dunfee, T. W. (1999) *Ties that bind: a social contracts approach to business ethics*, Boston, MA: Harvard Business School Press.

Draper, S. (2000) *Corporate Nirvana: is the future socially responsible?*, London: Industrial Society.

Driver, C. and Thompson, G. (2002) 'Corporate governance and democracy: the stakeholder debate revisited', *Journal of Management and Governance*, 6, 111–130.

Eaglesham, J. (2007) 'Carrots and sticks needed for business, says Tories', *Financial Times*, January 16, 6.

Economist, The (2008a) 'Just good business', *The Economist*, 17 January 2008, 65.

Economist, The (2008b) 'Strange bedfellows', *The Economist*, 24 May 2008, 99.

Edelman (2007a) *2007 Edelman Stakeholder Study: trust at the crossroads in Australia*, Sydney: Edelman.

Edelman (2007b) Sydney Morning Herald, 20 September, p.31.

Elkington, J. (1999) *Cannibals with forks: the triple bottom line of twenty first century business*, Oxford: Capstone Press.

Engels, F. (1884) *The origin of the family, private property and the State* [online]. Available at: http://www.marxists.org/archive/marx/works/1884/origin-family/index.htm

Farah, D. (2001) *The Washington Post*, November 20, 22.

Farjoun, M. (2002) 'Towards an organic perspective on strategy', *Strategic Management Journal*, 23, 561–594.

Fewings, P. (2006) 'The application of professional and ethical codes in the construction industry: a managerial view', *International Journal of Technology, Knowledge and Society*, 2 (7), 141–150.

Fewings, P. (2009) *Ethics for the built environment*, London: Taylor & Francis.

Fleishman-Hillard (2009) *Rethinking corporate social responsibility, Fleishman-Hillard Inc and National Consumers League Study* [online]. Available at: http://www.onlinefreebooks.net/social-culture-ebooks/society/rethinking-corporate-social-responsibility-a-fleishman-hillard-national-consumers-league-study-pdf.html

Fox, M. D. and Heller, M. A. (2006) 'What is good corporate governance', in Fox, M. D. and Heller, M. A. (eds) *Corporate governance lessons from transition economy reforms,* Princeton, NJ: Princeton University Press.

Frederick, W. C. (1998) 'Creatures, corporations, communities, chaos complexity', *Business and society*, 37 (12), 358–389.

Frederick, W. C. (2006) 'Pragmatism, nature and norms', *Business and Society Review*, 105 (4), 467–479.

Freeman, R. E. (1984) *Strategic management: a stakeholder approach*, Boston: Pitman.

Friedman, M. (1962) *Capitalism and freedom*, Chicago, IL: University of Chicago Press.

Galbraith, J. K. (2006) 'Does primacy stakeholder management positively affect the bottom line?', *Management Decision*, 44 (8), 1106–1121.

Gao, R. F. and Antolin, M. N. (2004) 'Stakeholder salience in corporate environmental strategy', *Corporate Governance*, 4, 65–76.

Ghoshal, S. (2005) 'Bad management theories are destroying management practice', *Academy of Management Learning and Education*, 4 (1), 75–91.

Giacalone, R. A., Jurkiewicz, C. L. and Deckop, J. R. (2008) 'On ethics and social responsibility: the impact of materialism, post materialism, and hope', *Human Relations*, 61, 483–514.

Giles, R. V. (1992) *Final Report by R.V. Gyles QC, Vols. 1–10*, Sydney: Royal Commission into Productivity in the Building Industry.

Gray, R. (2002) 'The social accounting project and accounting organisations and society: privileging engagement, imaginings, new accountings and pragmatism over critique', *Accounting Organisations and Society*, 27, 687–708.

Green, S. D. (2009) 'The evolution of corporate social responsibility in construction', in Murray, M. and Dainty, A. (eds) *Corporate social responsibility in the construction industry,* London: Taylor & Francis.

Green, S. D., Larsen, G. D. and Kao, C. C. (2008) 'Competitive strategy revisited: contested concepts and dynamic capabilities', *Construction management and economics*, 26, 63–78.

Greiner, L. E. (1972) 'Evolution and revolution as organisations grow', *Harvard Business Review*, July–August 1972, 37–46.

Hambrick, D. C. and Fredrickson, J. W. (2005) 'Are you sure you have a strategy', *Academy of Management Executive*, 19, 51–62.

Hammonds, K. H. (2008) 'Michael Porter's big ideas', *Fast Company*, 19 December [online]. Available at: http://www.fastcompany.com/magazine/44/porter.html

Hart, O. (1995) *Firms, contracts and financial structure*, Oxford: Clarendon Press.

Hartman, L. P. (1998) *Perspectives in business ethics*, London: McGraw-Hill.

Hawkins, P. (1997) 'Organisational culture: sailing between evangelism and complexity', *Human Relations*, 50, 417–440.

Heugens, P. and Dentchev, N. (2007) 'Taming Trojan horses: identifying and mitigating corporate social responsibility risks', *Journal of Business Ethics*, 75, 151–170.

Hill, R. P. and Cassill, D. L. (2004) 'The naturological view of the corporation and its social responsibility: an extension of the Frederick Model of corporation-community relationships', *Business and Society Review*, 109 (3), 281–296.
Hillebrandt, P. M. and Cannon, J. (1990) *The modern construction firm*, London: Macmillan.
Hinkley, R. (2002) 'How corporate law inhibits social responsibility', *Business Ethics Magazine*, Spring 2002 [online]. Available at: http://www.coneinc.com/pages/pr_13.html
Ho, C. F. M. (2003) Ethics management in a construction organisation: employee attitudes to corporate code of ethics, Unpublished PhD Thesis, University of New South Wales.
Holt, G. and Edwards, D. (2005) 'Domestic builder selection in the UK housing repair and maintenance sector: a critique', *Journal of Construction Research*, 6 (1), 123–137.
Horrigan, B. (2005) 'Comparative corporate governance developments and the key ongoing challenges from Anglo-American perspectives', in Tully, S. (ed.) *Research handbook on corporate legal responsibility,* Cheltenham: Edward Elgar Press.
Hubbard, G. (2004) *Strategic management: thinking analysis and action*, Sydney: Pearson Prentice Hall.
Hubbard, G., Samuel, D., Heap, S. and Cocks, G. (2002) *The first XI: winning organisations in Australia*, Sydney: Wiley.
Humphries, D. (2007) 'Huge deals of secrecy fuel skepticism', *The Sydney Morning Herald*, 6 March, 5.
Inglehart, R. (1977) *The silent revolution: changing values and political styles among Western publics*, Princeton, NJ: Princeton University Press.
Jacobs, D. C. (2004) 'A pragmatist approach to integrity in business ethics', *Journal of Management Inquiry*, 13 (3), 215–223.
Jensen, M. C. (2001) 'Value maximization, stakeholder theory, and the corporate objective function', *European Financial Management*, 7 (3), 297–317.
Jones, P., Comfort, D. and Hillier, D. (2006) 'Corporate social responsibility and the UK construction industry', *Journal of Corporate Real Estate*, 8 (3), 134–150.
Jopson, B. (2007) 'Why corporate reporting so seldom enlightens', *Financial Times*, 16 January, 13.
Kant, I. (1785) *Fundamental principles of the metaphysic of morals* (translated by Thomas Kingsmill Abbott) [online]. Available at: http://philosophy.eserver.org/kant/metaphys-of-morals.txt
Kaplan, R. S. and Norton, D. P. (1992) 'The balanced score card: measures that drive performance', *Harvard Business Review*, 70 (1), 71–79.
Kaptein, M. and Wempe, J. (2002) *The balanced company: a corporate integrity theory*, 1st edition, New York: Oxford University Press, 17, 250–256.
Kasperson, R., Jhaveri, N. and Kasperson, J. X. (2001) 'Stigma and the social amplification of risk: toward a framework of analysis', in Flynn, J., Slovic, P. and Kunreuther, H. (eds) *Risk, media and stigma: understanding public challenges to modern science and technology,* London and Sterling, VA: Earthscan Publications Ltd.
Kelly, G., Kelly, D. and Gamble, A. (1997) *Stakeholder capitalism*, London: Macmillan.
Kline, J. M. (2005) *Ethics for international business – decision making in a global political economy*, London and New York: Routledge.

Kotler, P. and Lee, N. (2005) *Corporate social responsibility*, Hoboken: John Wiley and Sons.

KPMG (2008) *KPMG international survey of corporate responsibility reporting 2008*, KPMG International Cooperative ('KPMG International') [online]. Available at: http://www.kpmg.com/Global/en/IssuesAndInsights/ArticlesPublications/Pages/Sustainability-corporate-responsibility-reporting-2008.aspx

Langford, D. and Male, S. (2001) *Strategic management in construction*, 2nd edition, Oxford: Blackwell Science.

La Porta, R. (1999) 'Corporate ownership around the world', *Journal of Finance*, 54 (2), 471–520.

Lehman, G. (2002) 'Global accountability and sustainability: research prospects', *Accounting Forum*, 26 (3), 219–232.

Lenin, V. I. (1917) *Imperialism, the highest stage of capitalism*, Moscow: Progress Publishers.

Lenzen, M. (2001) 'A generalized input-output multiplier calculus for Australia', *Economic Systems Research*, 13 (1), 65–92.

Lesser, E. L. (2000) *Knowledge and social capital*, Oxford: Butterworth-Heinemann.

Lingard, H., Blismas, N. and Stewart, P. (2009) 'CSR in the Australian Construction Industry', in Murray, M. and Dainty, A. (eds) *Corporate social responsibility in the construction industry*, London: Taylor & Francis.

Longshaw, P., Roper, A., May, B., Hastings, M. and Patterson, L. (2005) 'What I've learnt as a CSR practitioner', *Corporate Responsibility Management*, 1 (5), 34–38.

Loosemore, M., Dainty, A. and Lingard, H. (2003) *Human resource management in construction projects – strategic and operational aspects*, London: Taylor & Francis.

Lopez, M. V., Garcia, A. and Rodriguez, L. (2007) 'Sustainable development and corporate performance: a study based on the Dow Jones sustainability index', *Journal of Business Ethics*, 75, 285–300.

Louisot, J. (2009) 'Delegates see red over black times', *Risk Management Magazine*, 18–19.

Macfarlane, I. (2008) 'Look beyond greed – a debt ridden world of leverage made this mess', *Sydney Morning Herald*, 4 December 2008, 15.

Mak, S. L. (1999) 'Where are construction materials headed', *Building Innovation and Construction Technology*, August (8), 34–38.

Makadok, R. and Coff, R. (2009) 'Both market and hierarchy: an incentive-system theory of hybrid governance forms', *Academy of Management Review*, 34, 297–320.

Margolis, J. and Walsh, J. (2003) 'Misery loves companies: rethinking social initiatives by business', *Administrative Science Quarterly*, 48, 268–305.

Marris, R. (1964) *The economic theory of managerial capitalism*, London: Macmillan.

Marx, K. (1867) 'Capital: a critique of political economy', Volume 1, Moscow: Progress Publishers, as cited in Eastman, M. (1932) *Capital and other writings by Karl Marx*, New York: Modern Library Books.

Matheson, D. (2004) *The complete guide to good governance in organisations*, Auckland: Profile Books.

Matten, D. and Moon, J. (2008) '"Implicit" and "Explicit" CSR: a conceptual framework for a comparative understanding of corporate social responsibility', *Academy of Management Review*, 33, 404–425.

Mcintosh, M. and Thomas, R. (2004) 'Learning from company engagement with the global compact', in McIntosh, M., Waddock, S. and Kell, G. (eds) *Learning to talk: corporate citizenship and the development of the UN Global Compact*, Sheffield: Greenleaf.

McNulty, T., Roberts, J. and Stiles, P. (2005) 'Undertaking governance reform and research: further reflections on the Higgs Review', *British Journal of Management*, 16, 99–107.

Mertus, J. (2000) 'Considering non state actors in the new millennium: toward an expanded participation in norm generation and norm application', *New York University Journal of International Law and Politics*, 32 (2), 537–566.

Micklethwait, J. and Wooldridge, A. (2003) *The company: a short history of a revolutionary idea*, New York: Modern Library Books.

Miles, R. E. and Snow, C. C. (1978) *Organisational strategy, structure and process*, New York: McGraw-Hill.

Mintzberg, H. (1994) *The risk and fall of strategic planning*, London: Prentice Hall.

Mitchell, R. K., Agle, B. R. and Wood, D. J. (1997) 'Toward a theory of stakeholder identification and salience: defining the principle of who and what really counts', *Academy of Management Review*, 22, 853–886.

Moodley, K. and Preece, C. (2009) 'Community interaction in construction', in Murray, M. and Dainty, A. (eds) *Corporate social responsibility in the construction industry*, London: Taylor & Francis.

Morck, R. K. and Steier, L. (2007) 'The global history of corporate governance – an introduction', in Morck, R. K. (ed.) *A history of corporate governance around the world*, Chicago: National Bureau of Economic Research and The University of Chicago Press.

Narver, J. C. and Slater, S. F. (1990) 'The effect of a market orientation on business profitability', *Journal of Marketing*, 54, 20–35.

Nishtar, S. (2004) 'Public private partnerships in health – a global call to action', *Health Research Policy and System*, 2 (5), 56–57.

Ofori, G. (2007) 'Millennium development goals and construction: a research agenda', proceedings of the CME 25 Construction Management and Economics: Past, Present and Future Conference, Reading University, 16–18 July 2007, 78–89.

Ogbonna, E. and Harris, L. C. (1998) 'Managing organisational change: compliance or genuine change?', *British Journal of Management*, 9, 273–288.

Organization for Economic Cooperation and Development (2004) 'OECD principles of corporate governance' [online]. Available at: http.www.oecd.org/DATAOECD/32/18/31557724.pdf

Oury, J. (2007) *A Guide to Corporate Social Responsibility*, London: British Standards Institute (BSI).

Paine, L. S. (1994) 'Managing for organisational integrity', *Harvard Business Review*, March/April 1994, 106–117.

Pascale, R. (1984) 'Perspectives on strategy: the real story behind Honda's success', *California Management Review*, 26, 47–72.

Pearce, D. (2006) 'Is the construction sector sustainable? Definitions and reflections', *Building Research and Information*, 34, 3, 201–207.

Pemberton, S. (2005) 'Moral indifference and corporate manslaughter: compromising safety in the name of profit', in Tully, S. (ed.) *Research handbook on corporate legal responsibility*, Cheltenham: Edward Elgar Press.

Perrini, F. and Marino, A. (2006) 'The basis for launching a new social entrepreneurial venture', in Perrini, F. (ed.) *The new social entrepreneurship*, Cheltenham: Edward Elgar Press.

Perrini, F., Tencati, A. and Pivato, G. (2006) 'Sustainability and stakeholder management: the need for new corporate performance and evaluation systems', *Business strategy and the environment*, 15, 296–308.

Petrovic-Lazarevic, S. (2008) 'The development of corporate social responsibility in the Australian construction industry', *Construction Management and Economics*, 26 (2), 93–101.

Pfeffer, J. (2005), 'Changing mental models: HR's most important task', *Human Resource Management*, 44, 123–128.

Philips, M. (2009) 'Last men standing', *Risk Management Magazine*, 14–15.

Phua, F. T. T., Loosemore, M., Dunn, K. and Ozguc, U. (2010) 'Barriers to integration and attitudes towards cultural diversity in the construction industry', in Barrett, P. *et al.* (eds) *Building a better world*, CIB World Congress, University of Salford, Salford, UK, May 10–13, 1411–1423.

Porter, M. E. (1980) *Competitive strategy: techniques for analyzing industries and competitors*, New York: Free Press.

Porter, M. E. (1985) *Competitive advantage*, New York: Free Press.

Porter, M. E. (1987) *Corporate strategy: the state of strategic thinking*, New York: Free Press.

Posner, R. (1993) *The problems of jurisprudence*, Cambridge, MA: Harvard University Press.

Prahalad, C. K. and Hamel, G. (1990) 'The core competence of the corporation', *Harvard Business Review*, 90, 79–91.

Pye, A. and Pettigrew, A. (2005) 'Studying board context, process and dynamics: some challenges for the future', *British Journal of Management*, 16, 27–38.

Rasche, A. and Esser, D. E. (2007) 'Managing for compliance and integrity in practice', in Carter, C., Clegg, S., Kornberger, M., Laske, S. and Messner, M. (eds) *Business ethics as practice*, Cheltenham: Edward Elgar Press.

Ray, R. S., Hornibrook, J., Skitmore, M., and Fraser, A. Z. (1997) 'Ethics in tendering: a survey of Australian opinion and practice', *Construction Management and Economics*, 17, 139–153.

RICS (2008) 'Towards a low carbon built environment: a road map for action'. *RICS Research Series*, London: RICS.

Risk (2009) *Employees vital to best practice CSR*, RiskMagazine, Dec/Jan, Issue 1, 22–23.

Robbins, M. and Smith, D. (2005) 'Managing risk for corporate governance', London: British Standards Institution.

Roberts, J., McNulty, T. and Stiles, P. (2005) 'Beyond agency conceptions of the work of the non-executive director: creating accountability in the boardroom', *British Journal of Management*, 16, 5–26.

Robertson, C. and Crittenden, W. (2003) 'Mapping moral philosophies: strategic implications for multinational firms', *Strategic Management Journal*, 24, 385–392.

Rumelt, R. P., Schendel, D. and Teece, D. (1994) *Fundamental issues in strategy*, Boston, MA: Harvard Business School Press.

Saha, M. and Darnton, G. (2005) 'Green companies or green con-parties: are companies really green, or are they pretending to be?', *Business and Society Review*, 102 (2), 117–157.

Schein, E. H. (1984) 'Coming to a new awareness of organisational culture', *Sloan Management Review*, 25, 3–16.

Schwartz, H. S. (1990) *Narcissistic process and corporate decay*, New York: New York University Press.

Senge, P. (1990) *The fifth discipline*, New York: Harper and Row.

Sexton, E. (2009a) 'Dust to dust', *The Sydney Morning Herald*, 14–15 March 2009, 1.

Sexton, E. (2009b) 'Hardie board broke the law', *The Sydney Morning Herald*, 24 April, 1–2.

Sharpe, W. (2004) 'Talking points: managing stakeholder relations in PPP projects', *Public Infrastructure Bulletin*, March 2004, 8–15.

Shen, W. (2005) 'Improve board effectiveness: the need for incentives', *British Journal of Management*, 16, S81–S89.

Sikula, A. (1996) *Applied management ethics*, New York: Irwin.

Sims, R. R. and Brinkmann, J. (2003) 'Enron ethics (or, culture matters more than codes)', *Journal of Business Ethics*, 45, 243–256.

Singer, A. W. (2000) 'Can a company be too ethical?', in Richardson, J. E. (ed.) *Business ethics 00/01,* Guilford, CT: Dushkin/McGraw-Hill.

Sklair, L. (2002) *Globalization: capitalism and its alternatives*, 3rd edition, Oxford: Oxford University Press.

Smith, A. (1776) *The wealth of nations*, Harmondsworth: Penguin Books.

Smith, R. (1999) 'PFI: perfidious financial idiocy', *British Medical Journal*, 319, 2–3.

Smyth, D. (2006) 'Keynote address', in Fang, D. P., Choudry, R. M. and Hinze, J. W. (eds) *Global unity for safety and health in construction: proceedings of the CIB W99 International Conference*, Tsinghua University, Beijing.

Soule, E. (2002) 'Managerial moral strategies – in search of a few good principles', *Academy of Management Review*, 27, 114–124.

Spar, D. L. and La Mure, L. T. (2003) 'The power of activism: assessing the impact of NGOs on global business', *Californian Management Review*, 45 (3), 78–101.

Stansbury, N. (2005) *Exposing the foundations of corruption in corruption*, Berlin: Transparency International.

Stewart, T. (1997) *Intellectual capital: the wealth of organisations*, New York: Doubleday.

Stock, R. (2009) 'D&O in the new realm of risk', *Risk Magazine*, Issue 65, July, 14–16.

Suen, H., Cheung, S. and Mondejar, R. (2007) 'Managing ethical behavior in construction organisations in Asia: how the teachings of Confucianism, Taoism and Buddhism and globalization influence ethics management?', *International Journal of Project Management*, 25, 257–265.

Sullivan, R. (2005) 'The influence of NGOs on the normative framework for business and human rights', in Tully, S. (ed.) *International documents on corporate responsibility,* Cheltenham: Edward Elgar Press.

Sweet, R. (2009) 'Blame game', *ICON International Construction Review*, 2nd Quarter 2009, 10–11.

Taylor, F. W. (1911) *Principles of Scientific Management*, New York: Harper and Row.

Teece, D. J., Pisano, G. and Sheun, A. (1997) 'Dynamic capabilities and strategic management', *Strategic Management Journal*, 18, 537–553.

Teo, M. M. M. (2009) *An investigation of community based protest movement continuity against construction project*, Unpublished PhD Thesis, University of NSW.

Tosi, H. L., Shen, W. and Gentry, R. J. (2003) 'Why outsiders on boards can't solve the corporate governance problem', *Organisational Dynamics*, 32, 180–192.
Tow, D. and Loosemore, M. (2009) 'Corporate ethics in the construction and engineering industries', *ASCE journal of legal affairs and dispute resolution*, 1 (3), 122–130.
Transparency International (2005) *Transparency International's 2005 Global Corruption report*, Berlin: Transparency International.
Trevino, L. K., Weaver, G. R., Gibson, D. G. and Toffler, L. B. (1999) 'Managing ethics and legal compliance, what works and what hurts', *California Management Review*, 41 (2), 131–151.
Tully, S. (2005) *International documents on corporate responsibility*, Cheltenham: Edward Elgar Press.
UNEP (2009) United Nations Environment Programme [online]. Available at: http://www.unep.org
Van Oosterhout, J. P. H. and Kaptein, M. (2006) 'The internal morality of contracting: advancing the contractualist endeavor in business ethics', *Academy of Management Review*, 3, 521–539.
Vandekerckhove, W. (2007) 'Integrity: talking the walk instead of walking the talk', in Carter, C. S. C., Kornberger, M., Laske, S. and Messner, M. (eds) *Business ethics as practice*, Cheltenham: Edward Elgar Press.
Van Wijk, G. (2007) 'The paradoxical situation of ethics in business', in Djelic, M. and R. V. (eds) *Moral Foundations of Management*, Cheltenham: Edward Elgar Press.
Venkatraman, N. and Subramaniam, M. (2006) 'Theorizing the future of strategy: questions for shaping research in the knowledge economy', in Pettigrew, A., Thomas, H. and Whittington, R. (eds) *Handbook of strategy and management*, London: Sage.
Verhoef, G. (2007) 'Never mind the ethics, look at the price', *Building Magazine*, 23 November, 41.
Verrender, I. (2009) 'Governance is good, but tick-a-box tactics break the spirit', *The Sydney Morning Herald*, Weekend Business, October 3–4, 5.
Vilanova, M., Lozano, J. M. and Arenas, D. (2008) 'Exploring the nature of the relationship between CSR and competitiveness', *Journal of Business ethics*, 87, 1, 34–58.
Viljoen, J. and Dann, S. (2000) *Strategic management*, 3rd edition, Sydney: Longman Pearson Education Australia.
Vranceanu, R. (2007) 'The moral layer of contemporary economics: a virtue-ethics perspective', in Marie-Laure, D. and Vranceanu, R. (eds) *Moral foundations of management*, Cheltenham: Edward Elgar Press.
Washington, S. (2009) 'Reforms need to be assessed', *The Sydney Morning Herald*, 16 October, 6.
WBCSD (World Business Council for Sustainable Development) (1999) *Meeting changing expectations* [online]. Available at: http://www.wbcsd.org/DocRoot/hbdf19Txhmk3kDxBQDWW/CSRmeeting.pdf
Wearing, R. (2005) *Cases in corporate governance*, London: Sage.
Werther, W. B. and Chandler, D. (2006) *Strategic corporate social responsibility – stakeholders in a global environment*, London: Sage.
Westphal, J. D. and Kanna, P. (2003) 'Keeping directors in line: social distancing as a control mechanism in the corporate elite', *Administrative Science Quarterly*, 361–398.

Whetton, D. A., Rands, G. and Godfrey, P. (2006) 'What are the responsibilities of business to society', in Pettigrew, A., Thomas, H. and Whittington, R. (eds) *Handbook of strategy and management*, London: Sage.

Williams, R. (2009) 'When the truth is buried deliberately in the detail', *The Sydney Morning Herald: Weekend Business Edition*, 9–10 May 2009, 6–7.

World Urban Forum (2006) *Report of the third session of the World Urban Forum*, June 19–23, Vancouver: UN HABITAT.

Zarkada-Fraser, A. and Skitmore, M. (2000) 'Decisions with moral content: collusion', *Construction Management and Economics*, 18, 101–111.

Zawdie, G. and Murray, M. (2009) 'The role of construction and infrastructure development in the mitigation of poverty and disaster vulnerability in developing countries', in Murray, M. and Dainty, A. (eds) *Corporate social responsibility in the construction industry*, London: Taylor & Francis.

Zsolnai, L. (2006) 'Extended stakeholder theory', *Society and Business*, 1, 37–44.

Index

For Product Safety Concerns and Information please contact our EU
representative GPSR@taylorandfrancis.com
Taylor & Francis Verlag GmbH, Kaufingerstraße 24, 80331 München, Germany

www.ingramcontent.com/pod-product-compliance
Ingram Content Group UK Ltd.
Pitfield, Milton Keynes, MK11 3LW, UK
UKHW021507240926
475702UK00035B/612
9780415459099